Pioneering the Octonion Cosmos II: God-Space With a QED-like Hubble Expansion Model of Universes and Megaverses

Stephen Blaha Ph. D.
Blaha Research

A Million Dimension God-Space: Source of the Cosmos
Derivation of Octonion Spaces Cosmology
Interconnections of the Cosmos
Basis of Vector Meson Dominance
The Photon as a Universe Frozen in Time
Evidence for a Megaverse and a Maxiverse
Octonion Cosmology Embodies Color SU(4)
Blaha Universe Scale Factor: $a(t) = [(t + t_0)/t_{now}]^{g[1 + d/(t + t_0)] + ht}$
QED-like Vacuum-Polarization Growth of Universes/Megaverses
Big Bang t = 0 Interim Estimates: $a = 3.19 \times 10^{-93}$, $H = 3.58 \times 10^{218}$ km s^{-1}Mpc^{-1}
Reduction of QUeST, UTMOST, and the Maxiverse to Dimension-32 Atoms
Universe and Megaverse Molecules, Superverse Entity
Transformation of Octonion Space-time Dimensions to Real Space-time Dimensions
Splitting Symmetries into Factors Without Spontaneous Symmetry Breaking
Object-Oriented C++ View of Octonion Cosmology

Pingree-Hill Publishing
MMXXI

Rev. 00/00/01   April 15, 2021

To Margaret

Some Other Books by Stephen Blaha

All the Megaverse! Starships Exploring the Endless Universes of the Cosmos using the Baryonic Force (Blaha Research, Auburn, NH, 2014)

SuperCivilizations: Civilizations as Superorganisms (McMann-Fisher Publishing, Auburn, NH, 2010)

All the Universe! Faster Than Light Tachyon Quark Starships & Particle Accelerators with the LHC as a Prototype Starship Drive Scientific Edition (Pingree-Hill Publishing, Auburn, NH, 2011).

Unification of God Theory and Unified SuperStandard Model THIRD EDITION (Pingree Hill Publishing, Auburn, NH, 2018).

The Exact QED Calculation of the Fine Structure Constant Implies ALL 4D Universes have the Same Physics/Life Prospects (Pingree Hill Publishing, Auburn, NH, 2019).

Unified SuperStandard Theory and the SuperUniverse Model: The Foundation of Science (Pingree Hill Publishing, Auburn, NH, 2018).

Quaternion Unified SuperStandard Theory (The QUeST) and Megaverse Octonion SuperStandard Theory (MOST) (Pingree Hill Publishing, Auburn, NH, 2020).

Unified SuperStandard Theories for Quaternion Universes & The Octonion Megaverse (Pingree Hill Publishing, Auburn, NH, 2020).

The Essence of Eternity: Quaternion & Octonion SuperStandard Theories (Pingree Hill Publishing, Auburn, NH, 2020).

A Very Conscious Universe (Pingree Hill Publishing, Auburn, NH, 2020).

From Octonion Cosmology to the Unified SuperStandard Theory of Particles (Pingree Hill Publishing, Auburn, NH, 2020).

Available on Amazon.com, bn.com Amazon.co.uk and other international web sites as well as at better bookstores (through Ingram Distributors).

CONTENTS

FIGURES and TABLES

Introduction

This volume is Part II of *Pioneering the Cosmos*. It extends the presentation of features of Octonion Cosmology to make it a complete theory of the Cosmos. The beginning is the million dimension God-Space, from which the other octonion spaces (forming an octet), and their instances (particles), are derived.

The spaces are shown to be atomic in the senses that they are all composed of lattices (honeycombs) of a set of symmetries that we call Dimension-32 Atoms. Each atom has a corresponding set of fermions that we call Fermion-32 Atoms. These atoms contain an SU(4) (or an SU(3)⊗U(1)) symmetry. Octonion Cosmology naturally contains SU(4).

The initial design of each octonion space had an octonionic space-time. In this book we transform octonionic space-times to real space-times plus a set of symmetries that interrelate the parts of fermion spectrums. We present the new connecting symmetries for a QUeST universe, an UTMOST Megaverse, and the Maxiverse. In the case of a QUeST universe these (broken) symmetries relate "Normal" and Dark fermion sectors. We embody these symmetries in an augmented Unified SuperStandard Theory (UST) that we call NEWUST. (UST was derived using Logic previously.)

We show that the conventional form of symmetry breaking via the Higgs Mechanism is not sufficient. We introduce a new form of symmetry "splitting into factors" based on two types of inheritance.

Octonion Cosmology supports hierarchies of nested space instances. We consider an example of these hierarchies. It clearly illustrates an analogy between Octonion Cosmology and an Object-Oriented C++ program. The Cosmos may be such a "program."

Instances of octonion spaces, such as our universe, are "particles" that expand with time. This book presents a new formula for scale factors that describes expansion from the Big Bang to the present. It is based on a QED-like vacuum polarization that has similar parameters with the author's successful calculation of the Fine Structure Constant α.

A view of the photon is presented showing that it is analogous to a frozen universe. Part of the photon "universe" connects to the ElectroWeak interactions; part of the photon universe connects to the Strong interaction and supports a ρ Vector Meson Dominance (VDM) connection.

1. Octonion Octets and Decimets

The search for deeper roots of the Standard Model of Elementary Particles led us to derive a Unified SuperStandard Theory (UST) with the Standard Model as a subset. The derivation of the UST was based on an approach similar to that of Geometry. It started with a set of axioms and proceeded systematically to derive features. The features, that were found, were a superset of the Standard Model including a Dark Matter sector, four layers of particles and interactions, and additional interactions.

A significant similarity[1] between the symmetries of the Standard Model and subgroups of the Lorentz group led us to explore the possibility of a further unification of internal symmetries and space-time symmetries within the framework of a higher dimension theory.

Our search (in books in 2020—See References) led us to successfully base UST on Quaternion[2] Unified SuperStandard Theory (QUeST) with a remarkable match between the internal symmetries, interactions, particle spectrum, and space-time symmetry of both theories.

We described[3] a Cosmology based on a spectrum of octonion spaces that included interlocked universe and Megaverse instances[4] (particles) of two octonion subspaces within the Maxiverse octonion space with ten space-time dimensions. This Cosmology appears to be at the deepest level of physical reality.

In this chapter we describe additional features of the octonion spaces in greater detail..

1.1 Fundamental Axioms

In Blaha (2021a) we started with two fundamental axioms with the goal of creating a unified theory of internal symmetries and space-time.

*In the study of octonion spaces we did **not** use octonion algebra. We treated octonions as determining the specific dimensions of octonion spaces.*

We assume

---

[1] Blaha (2018e) and (2020c) as well as other books by the author.

[2] In Blaha (2020l) we established QUeST in an octonion framework.

[3] See Blaha (2020l).

[4] An instance is a "particle" of a space. The particle contains the space-time, fermions, bosons, and the consequences of interactions, which are described by the internal symmetries of the space. It differs from an elementary particle by supporting a growing size, and evolution, as well as having dimensions and contents. Our instances originate in highly energetic, far off-shell fermion-antifermion annihilation. Elementary particles are created in on-shell interactions.

1. An octet (octonion) of octonion spaces.

2. The spaces' dimensions progressively grow by factors of four from space to space with the smallest space initially having the dimensions of eight octonion coordinates where each coordinate is an octonion. Thus the number of dimensions in the smallest space is $8 \cdot 8 = 64$.

3. Each space has a square array of dimensions.

We numbered the components of the spaces octet from 0 through 7 and call each component a *level*.[5] The effect of axiom 2 is to make the total number of dimensions of level j to be four times that of level $j + 1$ for $j = 0, 1, \ldots 6$.

Axiom 3 follows naturally from the creation rules for octonion space instances from fermion-antifermion annihilation as described in chapters 3, 7, and 10 of Blaha (2021a).

In units of octonions the number of octonions *per coordinate* for levels 7 through 0 is 1, 2, 4, 8, 16, 32, 64, and 128. For level j the number of octonions in a coordinate is 2^{7-j} and the number of coordinates also is 2^{7-j}. The total number of dimensions is $2^{14-2j+6} = 2^{20-2j}$ since an octonion has 2^3 dimensions within it.

For example, each level 0 coordinate contains $128 \cdot 8 = 1024$ elements, and there are 1024 coordinates, with the result that the level 0 space has 1,048,576 dimensions. The above axioms imply Fig. 1.4.

1.2 Instances of the Octonion Spaces

The spaces defined in Fig. 1.1 can have instances (particles) created. Since we anticipate the existence of a space for universes and for Megaverses (Multiverses), we see a need to create universes instances (particles) within Megaverse (particles.) In chapters 3, 7, and 10 of Blaha (2021a) we described particle creation mechanisms based on the annihilation of a fermion-antifermion pair in a parent instance. Thus an instance of space 4 generates an instance of space 5 (a Megaverse) through the annihilation of a fermion-antifermion pair.

An instance of space 5 generates an instance of space 6 (a universe) through the annihilation of a fermion-antifermion pair. And so on. We found the features of the created instance are determined by the structure of the annihilating fermion-antifermion pair.

1.3 Nesting of Octonion Spaces

The spaces in Fig. 1.1 can be viewed as nested:

---

[5] The numbering is similar to that of Dante's *Inferno*. The numbers begin at 0, the highest level, and increase with descending levels to the lowest level. The numbering becomes more acceptable as we consider the roles of the space levels later.

> Space 0 has four copies of space 1
> Space 1 has four copies of space 2
> Space 2 has four copies of space 3
> Space 3 has four copies of space 4
> Space 4 has four copies of space 5
> Space 5 has four copies of space 6
> Space 6 has four copies of space 7

OCTONION SPACES LEVELS

Level Number

	Coordinate Type	Number of Coordinates	Number of Dimensions
0	Complex Octonion Octonion Octonion (2^7)	Complex Octonion Octonion Octonion	1024×1024
1	Octonion Octonion Octonion (2^6)	Octonion Octonion Octonion	512×512
2	Quaternion Octonion Octonion (2^5)	Quaternion Octonion Octonion	256×256
3	Complex Octonion Octonion (2^4)	Complex Octonion Octonion	128×128
4	Octonion Octonion (2^3)	Octonion Octonion	64×64
5	Quaternion Octonion[6] (2^2)	Quaternion Octonion	32×32
6	Complex Octonion[7] (2^1)	Complex Octonion	16×16
7	Octonion (2^0)	Octonion	8×8

Figure 1.1. The levels and dimensions of the octonion space octet. Note the lowest level is octonionic as it should be.

1.4 Match of Level 6 Space with UST and Level 5 with UTMOST

We identified the level 6 space instance with our universe's UST. We note that level 6 has 256 dimensions. An examination of the dimensions of the internal symmetry and space-time of the UST gives 256 dimensions as well – if the four dimensions of UST are generalized to octonion dimensions. Chapter 5 of Blaha (2021a) showed this match in detail. Therefore we take level 6 space to be the space of our universe.

Based on this choice we can identify an instance of space 5 as our Megaverse. (Other Megaverse instances are possible.) We can further identify a space 4 instance as the "parent" of space 5 instances. We will assume (and show later in chapter 7) that there is only one instance of space 4 leading to our universe. We call that instance the Maxiverse.

---

[6] In our earlier books in 2020 we also designated this 1024 dimension space (5) as 64 complex octonion space.

[7] In our earlier books in 2020 we also designated this 256 dimension space(6) as 32 complex quaternion space.

1.5 Octonion Space 7

Octet space 7 presented a problem in that we cannot derive it as following from a fermion-antifermion annihilation in level 6. We then showed that such a process generates an instance of a 4 × 4 = 16 dimension space. That space in turn can generate another 4 × 4 = 16 dimension space, which can in turn generate yet another 4 × 4 = 16 dimension space, and another. We can truncate the generation of such spaces to yield four spaces in total. The resulting pattern of fermion-antifermion annihilation generations appears in Fig. 1.2.

The four lowest spaces, if combined, give the 8 × 8 = 64 dimension space of Fig. 1.4. It contains a new octet of octonion spaces: new in the senses that spaces 7 through 10 of Fig. 1.3 are combined into space 7 of Fig. 1.4.

Thus we have an eight octonion space spectrum.

1.6 Implications of the New Eight Octonion Space Spectrum

Assembling the octet of octonion spaces in Fig. 1.4 one is led to consider the possibility of a deeper significance. We can expression the octet as an octonion expression:

$$O = S_1\mathbf{a} + S_2\mathbf{b} + \ldots + S_7\mathbf{h} \tag{1.1}$$

where **a**, **b**, … **h** are fundamental octonion units.

It evokes the impression that there is a deeper topological significance that justifies Octonion Cosmology at a yet deeper level.

There are clearly two possibilities for justification. One is a deeper topological theory. The other is a justification is exclusion of all cases where no other alternative is possible without a consequent disaster such as divergences in physically measurable quantities. Currently, we have no evidence for either alternative.

1.7 Symmetry of the Eight Space Spectrum

The new eight space spectrum has the *initial* symmetry:

$$U(524288) \otimes U(131072) \otimes U(32768) \otimes U(8192) \otimes U(2048) \otimes U(512) \otimes U(128) \otimes U(32) \otimes U(8) \tag{1.2}$$

We present a derivation of the eight and ten space spectrums in chapter 6.

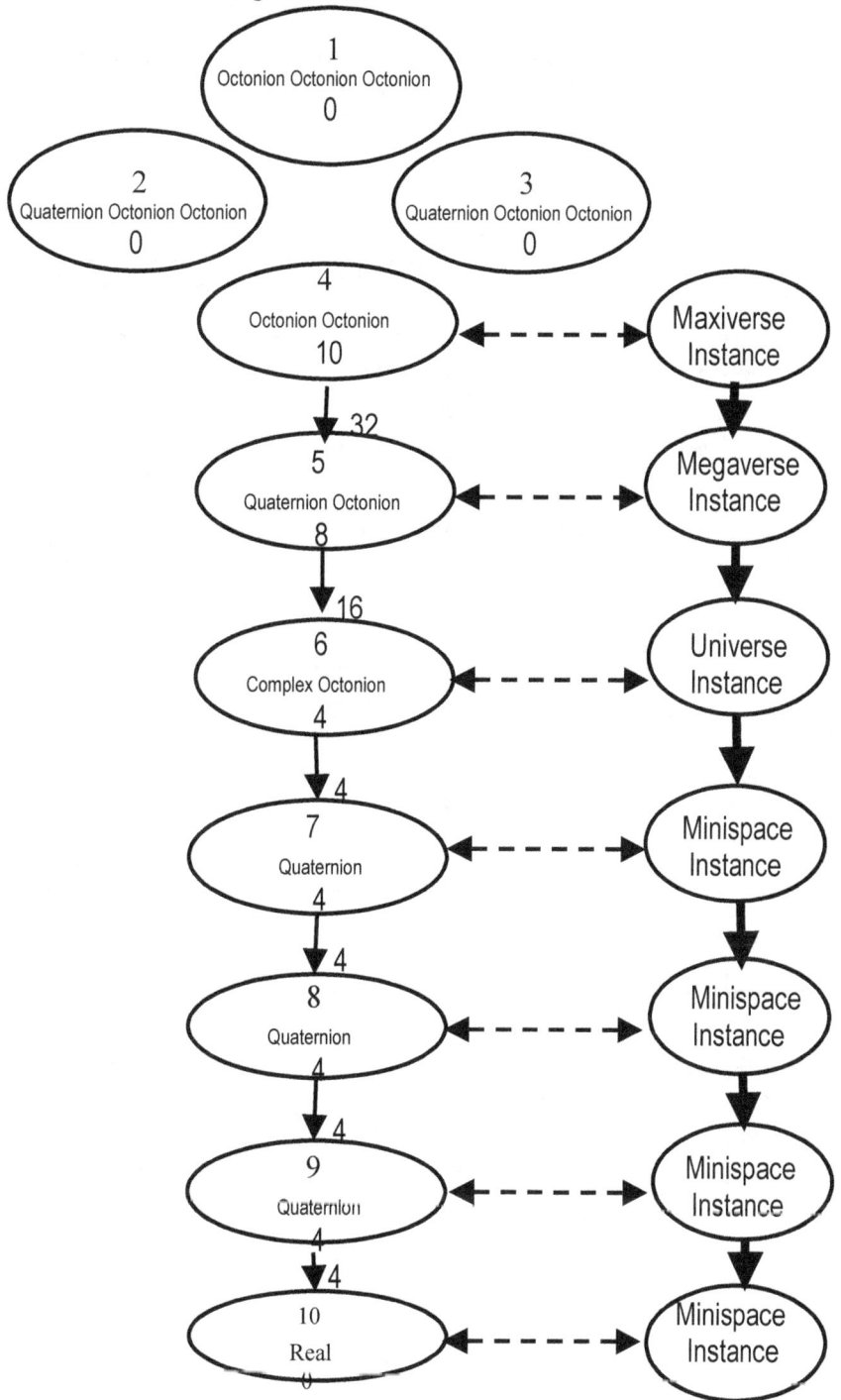

Figure 1.2. Chain of Instances. The ten spaces with their spectrum number and their space-time dimensions are indicated within each oval. The number of spinor components for each fermion - antifermion pair that annihilates to

produce the "next space down" is specified next to each arrow for the lower spaces. On the right are symbols for instances (particles) of spaces.

OCTONION SPACES SPECTRUM

Spectrum Number ▼

	Coordinate Type	Number of Coordinates	Dimension Array Size	Space-Time Dimensions
	Superverse Space			
0	Complex Octonion Octonion Octonion (1024)	Complex Octonion Octonion Octonion	1024×1024	0
	Spaceless[8]			
1	Octonion Octonion Octonion (512)	Octonion Octonion Octonion	512×512	0
2	Quaternion Octonion Octonion (256)	Quaternion Octonion Octonion	256×256	0
3	Complex Octonion Octonion (128)	Complex Octonion Octonion	128×128	0
	Cosmology[9]			
4	Octonion Octonion (64)	Octonion Octonion	64×64	10 quaternion octonion
	Maxiverse Space			
5	Quaternion Octonion[10] (32)	Quaternion Octonion	32×32	8 complex octonion
	Megaverses Space			
6	Complex Octonion[11] (16)	Complex Octonion	16×16	4 octonion
	Universe Space			
	Minispaces[12]			
7	Quaternion (4)	Quaternion	4×4	4 Real
8	Real (4)	Real (4)	4×4	4 Real
9	Real (4)	Real (4)	4×4	4 Real
10	Real (4)	Real (4)	4×4	0

Figure 1.3. The spectrum of the Superverse and the ten octonion spaces. The spaces are numbered from 0 through 10. The numbers in parentheses in column 2 are the number of dimensions in each coordinate. The items in column 3 are the number of rows of dimensions (1024, 512, 256, 128, 64, 32, 16, 4, 4, 4, 4).

---

[8] Spaceless spaces are spaces without a space-time.

[9] Cosmological spaces are those, which are directly related to physics from megaverses and universes to elementary particles.

[10] In our earlier books in 2020 we also designated this 1024 dimension space (5) as 64 complex octonion space.

[11] In our earlier books in 2020 we also designated this 256 dimension space(6) as 32 complex quaternion space.

[12] A Minispace is a subspace of a universe space.

"New" OCTONION SPACES SPECTRUM

Spectrum Number

	Coordinate Type	Number of Coordinates	Dimension Array Size	Space-Time Dimensions
0	**Superverse Space** Complex Octonion Octonion Octonion (1024)	Complex Octonion Octonion Octonion	1024×1024	0
	Spaceless[13]			
1	Octonion Octonion Octonion (512)	Octonion Octonion Octonion	512×512	0
2	Quaternion Octonion Octonion (256)	Quaternion Octonion Octonion	256×256	0
3	Complex Octonion Octonion (128)	Complex Octonion Octonion	128×128	0
	Cosmology[14]			
4	Octonion Octonion (64)	Octonion Octonion	64×64	10 quaternion octonion
	Maxiverse Space			
5	Quaternion Octonion[15] (32)	Quaternion Octonion	32×32	8 complex octonion
	Megaverses Space			
6	Complex Octonion[16] (16)	Complex Octonion	16×16	4 octonion
	Universe Space			
	Composite Minispace[17]			
7	Octonion (8) (Composite)	Octonion	8×8	2 Octonion

Figure 1.4. The spectrum of the Superverse and the eight octonion spaces. The spaces are numbered from 0 through 7. The numbers in parentheses in column 2 are the number of dimensions in each coordinate. The items in column 3 are the number of rows of dimensions – coordinates. (1024, 512, 256, 128, 64, 32, 16, 8).

---

[13] Spaceless spaces are spaces without a space-time.

[14] Cosmological spaces are those, which are directly related to physics from megaverses and universes to elementary particles.

[15] In our earlier books in 2020 we also designated this 1024 dimension space (5) as 64 complex octonion space.

[16] In our earlier books in 2020 we also designated this 256 dimension space(6) as 32 complex quaternion space.

[17] A Minispace is a subspace of a universe space.

2. Symmetry Splitting in Octonion Cosmology

There is a general belief that there exists an overall symmetry for an entire universe/Megaverse. Spontaneous symmetry breakdowns are viewed as the source of the separation of particle symmetries into SU(3), SU(2)⊗U(1), and so on, as well as the breaking of symmetries such as ElectroWeak SU(2)⊗U(1) breakdown.

We put forward the proposition that the presumed inherent U(128) symmetry of 256 dimension QUeST universes (and similarly for Megaverses and other spaces) never existed.[18] Universes and Megaverses *began* with factored symmetries from their moment of origin in a fermion-antifermion annihilation.

In chapter 6 we describe the splits generating the 10 octonion spaces, which we also believe existed from the first "moment" of existence of the Cosmos.

We thus see three types of symmetry splitting:

1. Splitting of the Superverse into 10 octonion spaces (or 8 if we combine the Minispaces.) We call this *global splitting by inheritance* since it generates entire octonion spaces.

2. Splitting of each octonion space into sets of Dimension-32 atoms as described in chapter 3. We call this type of splitting *local splitting by inheritance* since it splits the symmetry of an individual octonion space.

3. Splitting of a symmetry within a Dimension-32 atom such as SU(2)⊗U(1), SU(4), a U(4) Generation group, and a U(4) Layer group. This splitting is accomplished by spontaneous symmetry breaking.[19]

This chapter describes splitting of type 2. Chapter 6 describes splitting of type 1. Spontaneous symmetry breaking (type 3) is described in Blaha (2018e) and (2020c).

2.1 Splitting of Type 2 Local Splitting by Inheritance

We saw in Blaha (2021a) that the overall structure of particle symmetries is determined in the large by the spinor structure of the annihilating fermion-antifermion pairs that generate universes[20] and megaverses.[21] *Thus there is a splitting of symmetries*

---

[18] Chapter 6 suggests the Superverse is the origin of the ten octonion spaces. The ten spaces are shown as a "splitting" of the Superverse. As above, we propose that the split-generated factoring of spaces existed from the "Beginning" – not as a result of spontaneous symmetry breaking.

[19] At best, splitting of type 3 is the only splitting that may be relevant to running coupling constant estimates of the symmetry unification energy.

[20] Chapter 4 of Blaha (2021a).

into a product of factors that is not due to spontaneous symmetry breaking (as it is usually envisioned.[22])We call this type of splitting local splitting by inheritance since it is, in a very real sense, inherited from a "parent" octonion space instance.

2.2 QUeST Inheritance

The spinor structure of an annihilating fermion-antifermion pair in an UTMOST Megaverse causes 4×4 blocks of dimensions in the resulting QUeST universe dimension array that factors the overall symmetry into blocks of internal symmetries. These blocks are described in detail in chapter 3. The blocks are not the result of symmetry breaking but reflect the structure of the spinors in UTMOST Megaverse fermions.[23] The sixteen 16-spinors of an 8- dimension Megaverse fermion are depicted in Figs. 2.1 and 2.2.: Fig. 2.3 shows the resulting internal symmetry 4×4 block structure of QUeST. See chapter 3 for a detailed discussion of these blocks.

2.3 UTMOST Megaverse Inheritance

The spinor structure of an annihilating fermion-antifermion pair in the Maxiverse causes 8×8 blocks of dimensions in the resulting UTMOST Megaverse that determine blocks of internal symmetries.[24] These blocks are described in detail in chapter 8 of Blaha (2021a). The blocks are not the result of symmetry breaking but reflect the structure of the spinors in Maxiverse urfermions (fermions). The thirty-two 32-spinors of a 10 dimension Maxiverse urfermion are depicted in Fig. 2.4. Spinor parts map to 8×8 blocks of internal symmetries (Fig. 2.5) in the UTMOST Megaverse. See chapter 3 for a detailed discussion of these blocks.

2.4 QUeST – UST Map and Map to UTMOST

The 4×4 blocks of QUeST are naturally determined by a map from QUeST to the Unified SuperStandard Theory (UST) that is shown in chapter 3 of Blaha (2019a). *The composition of the 8×8 blocks in the UTMOST Megaverse is determined by mapping upward from QUeST since the UTMOST dimension array consists of four copies of QUeST.*

---

[21] Chapter 8 of Blaha (2021a).

[22] Spontaneous symmetry breaking does appear to take place in ElectroWeak Theory: SU(2)⊗U(1) breaking, and in the Generation and Layer groups, as well as the breaking of Strong groups from SU(4) to SU(3)⊗U(1).

[23] For that reason this form of symmetry factoring can be viewed as evidence for the existence of our universe in a Megaverse.

[24] The breakdown into 64 dimension blocks carries over to QUeST. Each layer of QUeST is a 64 dimension block.

Number of Columns = 4   4   4   4

	u-type fermion)	v-type (anti-fermion)	
4	u spin up	v small terms 1	
4	u spin down	v small terms 2	
4	u small terms 1	v spin down	
4	u small terms 2		v spin up

Figure 2.1. The 16 spinors of a 8 dimension spinor. Each spinor has 16 rows.

u-up–v-v1	u-up–v-v2	u-up–v-down	u-up–v-up
u-up–v-v1	u-up–v-v2	u-up–v-down	u-up–v-up
u-up–v-v1	u-up–v-v2	u-up–v-down	u-up–v-up
u-up–v-v1	u-up–v-v2	u-up–v-down	u-up–v-up
u-down–v-v1	u-down–v-v2	u-down–v-down	u-down–v-up
u-down–v-v1	u-down–v-v2	u-down–v-down	u-down–v-up
u-down–v-v1	u-down–v-v2	u-down–v-down	u-down–v-up
u-down–v-v1	u-down–v-v2	u-down–v-down	u-down–v-up
u-v1–v-v1	u-v1–v-v2	u-v1–v-down	u-v1–v-up
u-v1–v-v1	u-v1–v-v2	u-v1–v-down	u-v1–v-up
u-v1–v-v1	u-v1–v-v2	u-v1–v-down	u-v1–v-up
u-v1–v-v1	u-v1–v-v2	u-v1–v-down	u-v1–v-up
u-v2–v-v1	u-v2–v-v2	u-v2–v-down	u-v2 v up
u-v2–v-v1	u-v2–v-v2	u-v2–v-down	u-v2–v-up
u-v2–v-v1	u-v2–v-v2	u-v2–v-down	u-v2–v-up
u-v2–v-v1	u-v2–v-v2	u-v2–v-down	u-v2–v-up

Figure 2.2. Outer product array [U_{ba}] (eq. 1.9) of the composite u-type and v-type spinors illustrating the structure of the outer product array of uv's.

	4	4	4	4
4	u-up–v-v1	u-up–v-v2	u-up–v-down	u-up–v-up
4	u-down–v-v1	u-down–v-v2	u-down–v-down	u-down–v-up
4	u-v1–v-v1	u-v1–v-v2	u-v1–v-down	u-v1–v-up
4	u-v2–v-v1	u-v2–v-v2	u-v2–v-down	u-v2–v-up

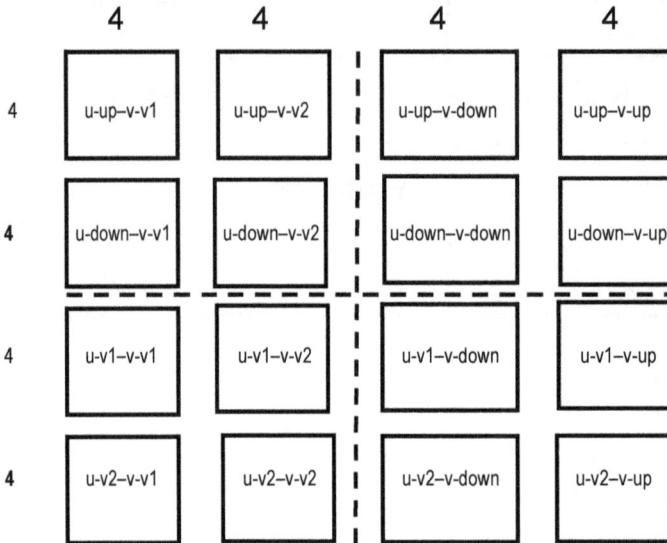

Figure 2.3. Block form of the 16 × 16 [U_{ba}] array. This is also the form of the QUeST dimension array of 256 dimensions. The blocks are divided by dashed lines that separate 64 dimension sections. These sections map to layers in QUeST (UST) as shown in Figs. C.2 and C.3.

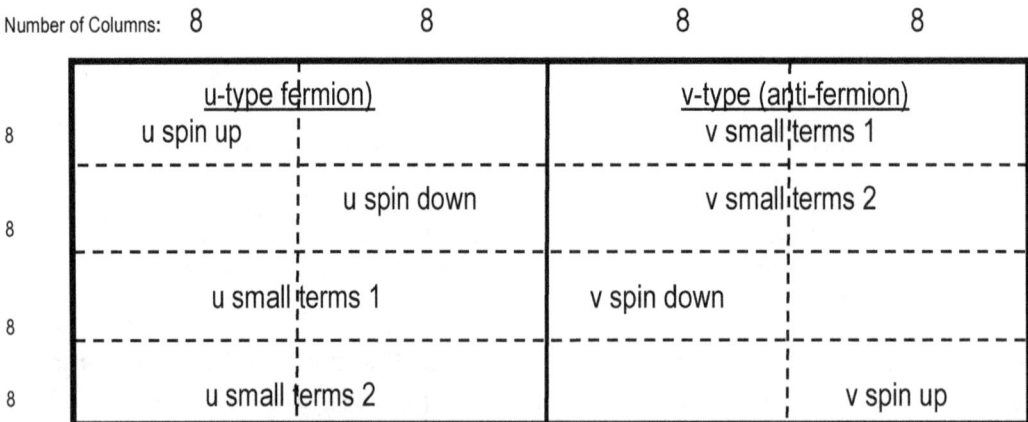

Number of Columns: 8   8   8   8

	u-type fermion)		v-type (anti-fermion)	
8	u spin up		v small terms 1	
8		u spin down	v small terms 2	
8	u small terms 1		v spin down	
8	u small terms 2			v spin up

Figure 2.4. The thirty-two 32-spinors of a 10 dimension urfermion (a Maxiverse fermion). Each spinor has 32 rows.

u-up–v-v1	u-up–v-v2	u-up–v-down	u-up–v-up
u-up–v-v1	u-up–v-v2	u-up–v-down	u-up–v-up
u-up–v-v1	u-up–v-v2	u-up–v-down	u-up–v-up
u-up–v-v1	u-up–v-v2	u-up–v-down	u-up–v-up
u-up–v-v1	u-up–v-v2	u-up–v-down	u-up–v-up
u-up–v-v1	u-up–v-v2	u-up–v-down	u-up–v-up
u-up–v-v1	u-up–v-v2	u-up–v-down	u-up–v-up
u-up–v-v1	u-up–v-v2	u-up–v-down	u-up–v-up
u-down–v-v1	u-down–v-v2	u-down–v-down	u-down–v-up
u-down–v-v1	u-down–v-v2	u-down–v-down	u-down–v-up
u-down–v-v1	u-down–v-v2	u-down–v-down	u-down–v-up
u-down–v-v1	u-down–v-v2	u-down–v-down	u-down–v-up
u-down–v-v1	u-down–v-v2	u-down–v-down	u-down–v-up
u-down–v-v1	u-down–v-v2	u-down–v-down	u-down–v-up
u-down–v-v1	u-down–v-v2	u-down–v-down	u-down–v-up
u-down–v-v1	u-down–v-v2	u-down–v-down	u-down–v-up
u-v1–v-v1	u-v1–v-v2	u-v1–v-down	u-v1–v-up
u-v1–v-v1	u-v1–v-v2	u-v1–v-down	u-v1–v-up
u-v1–v-v1	u-v1–v-v2	u-v1–v-down	u-v1–v-up
u-v1–v-v1	u-v1–v-v2	u-v1–v-down	u-v1–v-up
u-v1–v-v1	u-v1–v-v2	u-v1–v-down	u-v1–v-up
u-v1–v-v1	u-v1–v-v2	u-v1–v-down	u-v1–v-up
u-v1–v-v1	u-v1–v-v2	u-v1–v-down	u-v1–v-up
u-v1–v-v1	u-v1–v-v2	u-v1–v-down	u-v1–v-up
u-v2–v-v1	u-v2–v-v2	u-v2–v-down	u-v2–v-up
u-v2–v-v1	u-v2–v-v2	u-v2–v-down	u-v2–v-up
u-v2–v-v1	u-v2–v-v2	u-v2–v-down	u-v2–v-up
u-v2–v-v1	u-v2–v-v2	u-v2–v-down	u-v2–v-up
u-v2–v-v1	u-v2–v-v2	u-v2–v-down	u-v2–v-up
u-v2–v-v1	u-v2–v-v2	u-v2–v-down	u-v2–v-up
u-v2–v-v1	u-v2–v-v2	u-v2–v-down	u-v2–v-up
u-v2–v-v1	u-v2–v-v2	u-v2–v-down	u-v2–v-up

Figure 2.5. Outer 32 × 32 product array of the composite u-type and v-type spinors illustrating the structure of the outer product array of uv's. Each column represents 8 spinor columns, making 32 columns in all.

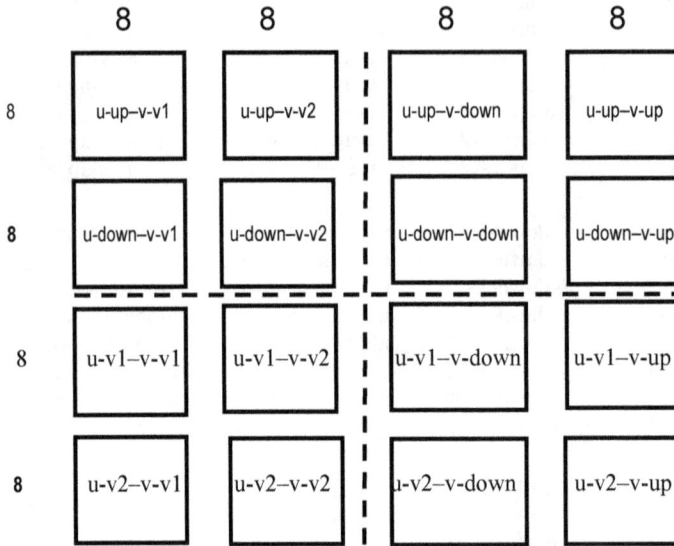

Figure 2.6. Block form of the resulting 32 × 32 array. This is also the form of the UTMOST dimension array of 1024 dimensions. It has 16 × 16 blocks, which have four 8 × 8 blocks within them. The set of four of 16 × 16 blocks above are divided by dashed lines into 256 dimension sections. These 256 dimension sections map to layers in UTMOST. Each UTMOST layer is equivalent to a QUeST 256 dimension array.

3. Atoms, Molecules, and an Entity

An examination of the structure of QUeST universes and UTMOST Megaverses shows that they are composed of sets of identical units having 32 dimensions composed of two 16 dimension parts. As chapter 2 showed these patterns are the result of their origin in fermion-antifermion annihilation, and in particular, of their spinor structures. In this chapter we analyze the content of octonion spaces and show they may be viewed as composed of honeycombs of "atoms" just as matter is composed of atoms.

To that end we define 16 dimension blocks, of type A and B below. We further define a *dimension-32 Atom* as consisting of a *pair* of 16 dimension blocks:

Dimension-32 Atom

$$\text{A.. } U(1) \otimes SU(2) \otimes U(1) \otimes SU(3) \otimes SL(2, C)^{25} \tag{3.1}$$
$$\text{B. } U(4) \otimes U(4)$$

Both "Normal" and Dark *dimension-32 Atoms* exist.

Their form is depicted in Fig. 3.1a and 3.1b. We call these pairs of 16 dimension blocks "atoms" because they are the constituents of the QUeST and UTMOST dimension arrays and symmetries.

The 256 dimensions of QUeST consist of eight dimension-32 atoms with two dimension-32 atoms in each of the four layers. See Fig. 3.2. They may be $SU(3) \otimes U(1)$ or $SU(4)$ dimension-32 atoms.

An SU(3)⊗U(1) Dimension-32 Atom

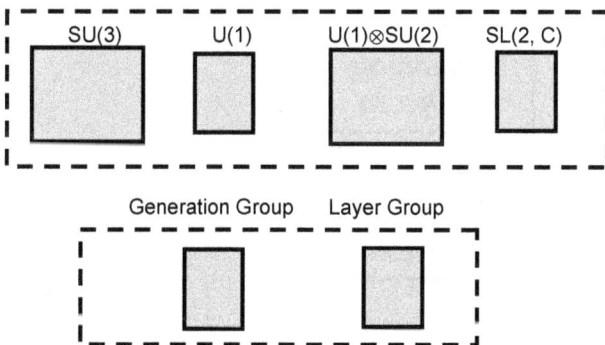

Figure 3.1a. An SU(3)⊗U(1) Dimension-32 atom consisting of a pair of 16 dimension blocks with each encircled by "dotted" lines..

---

[25] SL(2, **C**) represents an SO⁺(1,3) vector representation

An SU(4) Dimension-32 Atom

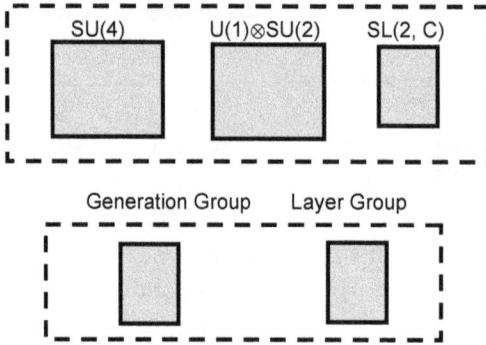

Figure 3.1b. An SU(4)-based, Dimension-32 atom consisting of a pair of 16 dimension blocks (with each encircled by "dotted" lines.).

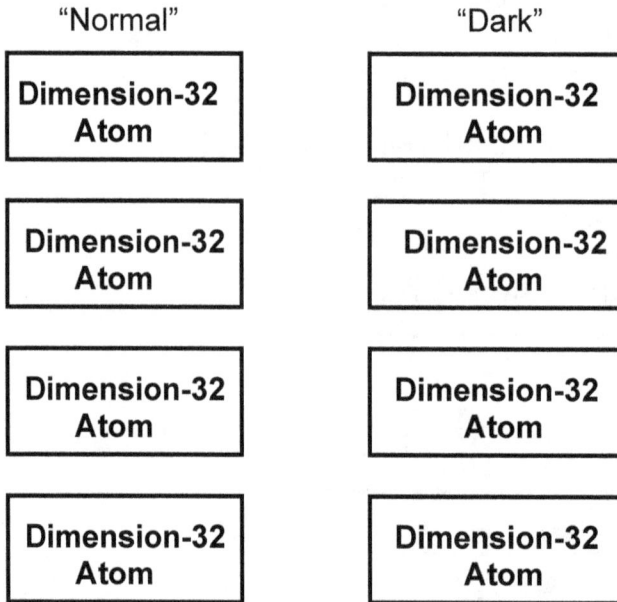

Figure 3.2. Eight Dimension-32 atoms arranged in the four layers of the 16 × 16 QUeST dimension array with each block having the form of Fig. 3.1a or 3.1b.

3.1 Features of Dimension-32 Atoms

We saw in Blaha (2019a), and earlier books, that the QUeST and UTMOST spaces are composed of dimension-32 atoms although we did not identify them as such. Each dimension-32 atom has its own set of internal symmetries[26]

$$SU(2) \otimes U(1) \otimes SU(4) \otimes U(4) \otimes U(4) \otimes SL(2,\mathbf{C}) \qquad (3.2)$$

or

$$SU(2) \otimes U(1) \otimes SU(3) \otimes U(1) \otimes U(4) \otimes U(4) \otimes SL(2,\mathbf{C}),$$

In eq. 3.1 we defined type A and type B parts of a dimension-32 atom in terms of its group content. We now note that the fundamental representation of the type A group has 16 dimensions, and the fundamental representation of the type B group also has 16 dimensions resulting in 32 dimensions for the fundamental representation of the dimension-32 atom.

Correspondingly, we find the sum of the vector boson fields of a dimension-32 atom is 32. Thus there is a certain unity in the dimension-32 atom.

From dimension-32 atom to dimension-32 atom we find the internal symmetries differ, in general, as do their coupling constants and gauge field masses.

3.2 Fermion-32 Atoms

Each dimension-32 atom has a corresponding set of fundamental fermions that combine (conceptually) to form a *Fermion-32 Atom*. As noted in earlier books, the total number of dimensions of QUeST (and UTMOST) equals the total number of fundamental fermions. For example, in the case of QUeST there are 256 dimensions and 256 fundamental fermions.

This equality holds down to the level of atoms. Each dimension-32 atom has 32 fundamental representation dimensions and a corresponding set of 32 fundamental fermions. We call the set of fermions a Fermion-32 Atom.

Fig. 3.3 gives the form of the 32 fermions in a fermion-32 atom. Note that four generations are required to have the number of dimensions of dimension-32 atoms equal the number of fundamental fermions (providing a justification for 4 generations.) There are "Normal" matter fermion-32 atoms as well as "Dark" matter fermion-32 atoms.

From fermion-32 atom to fermion-32 atom we find the internal symmetries differ, in general, as do their coupling constants and masses.

3.3 Atoms of Octonion Spaces

The eight/10 octonion spaces can be viewed as composed of dimension-32 atoms arranged in layers.

3.3.1 QUeST Spaces

A QUeST space has eight dimension-32 atoms arranged in four layers with a normal and Dark dimension-32 atom in each layer. There is a corresponding set of eight fermion-32 atoms. Fig. 3.4 shows the four QUeST layers of fermions. There are 8 fermion-32 atoms in QUeST arranged in four layers of normal and Dark dimension-32 fermions.

---

[26] The Layer groups straddle the layers. Each Layer group "rotates" fermions of the same generation in all four layers.

3.3.2 UTMOST Spaces

An UTMOST space has thirty-two dimension-32 atoms arranged in four layers with a "normal" and seven Dark dimension-32 atoms in each layer. There is a corresponding set of thirty-two fermion-32 atoms.

3.3.3 Other Spaces

The other spaces have dimension-32 atoms since we utilize upward mapping from QUeST and UTMOST remembering that there are four QUeSTs in UTMOST, four UTMOSTS in the Maxiverse and so on.

The number of dimension-32 atoms with a corresponding number of fermion-32 atoms is:

$$\begin{array}{lll}
\text{QUeST} & \text{8 dimension-32 atoms} & (3.3) \\
\text{UTMOST} & \text{32 dimension-32 atoms} & \\
\text{Maxiverse} & \text{128 dimension-32 atoms} & \\
\text{Space 3} & \text{512 dimension-32 atoms} & \\
\text{Space 2} & \text{2,048 dimension-32 atoms} & \\
\text{Space 1} & \text{8,192 dimension-32 atoms} & \\
\text{Space 0} & \text{32,768 dimension-32 atoms} &
\end{array}$$

3.4 Universe and Megaverse Molecules

Having defined atoms of dimensions and fermions it seems natural to view the various spaces as being forms of molecules. One often hears references to the universe as alive. This concept accords well with these atoms of dimensions and fermions.

3.5 A Superverse Entity

Space 0 which we call the Superverse (or God-Space) has 32,768 dimension-32 atoms. It also contains the other spaces as subsets. For example, it has 4,096 QUeST universe spaces within it. Based on the concepts of atoms and molecules we can view a Superverse instance as an "Entity." In chapter 6 we shall see that the entity spawns the other spaces using an inheritance mechanism.

8

e	v	uq1	uq2	uq3	dq1	dq2	dq3
e	v	uq1	uq2	uq3	dq1	dq2	dq3
e	v	uq1	uq2	uq3	dq1	dq2	dq3
e	v	uq1	uq2	uq3	dq1	dq2	dq3

4

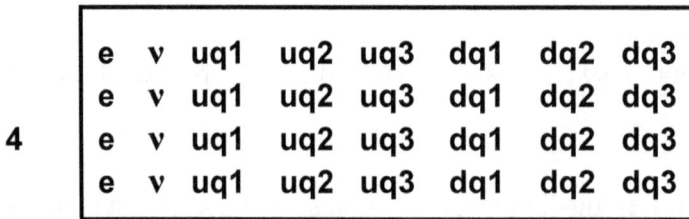

Figure 3.3. A fermion-32 atom consisting of four generations of fermions. The result is an 8 × 4 block. The identifier "uq" signifies an up-type quark. The identifier "dq" signifies a down-type quark. Both "Normal" and "Dark" sets of fermions are fermion-32 atoms. Thus each level of QUeST fermions consists of two fermion-32 atoms.

"Normal" 8	"Dark" 8
4 Fermion-32 Atom	Fermion-32 Atom
4 Fermion-32 Atom	Fermion-32 Atom
4 Fermion-32 Atom	Fermion-32 Atom
4 Fermion-32 Atom	Fermion-32 Atom

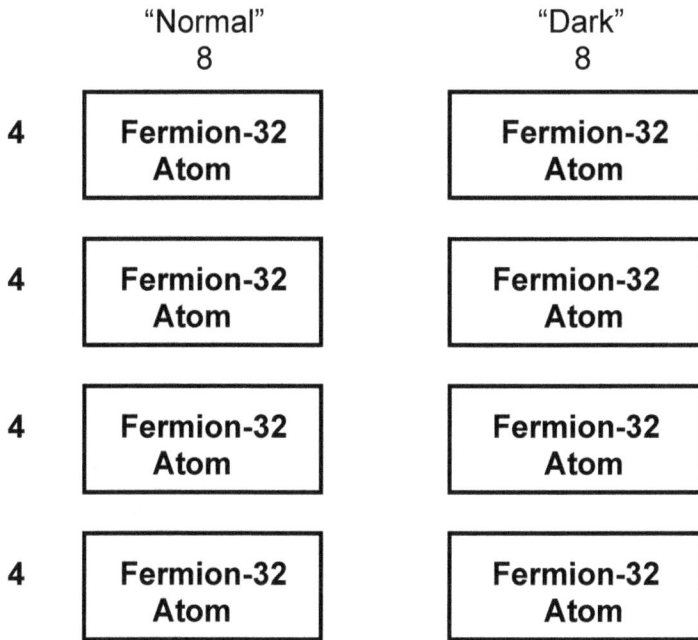

Figure 3.4. Fermion-32 atoms in the 16 × 16 QUeST fermion array with each row corresponding to one layer. Each block contains four generations of fermions as in Fig. 3.3.. The result is 4 × 8 blocks. We see "Normal" and "Dark" fermion-32 atoms in each layer.

4. Space-Time Dimensions vs. Internal Symmetry Dimensions

In QUeST and UTMOST we found a four octonion space-time and an eight complex octonion space-time respectively. In UST we specified a real four dimension space-time. We now propose real four dimension QUeST and real eight dimension UTMOST by extending the set of internal symmetries to absorb the octonion parts of their dimensions. The transfer of dimensions[27] from space-times to internal symmetries enables us to define a new set of symmetries[28] that supports new interactions between "normal" and "Dark" matter. *We choose NOT to have "unused" dimensions in octonion spaces.*

The new interactions result in more interconnected QUeST universes, UTMOST Megaverses, and other octonion space instances. In each space-time instance every fermion is "connected" to every other fermion through a chain of interactions. There are no truly isolated fermions or sets of fermions. This connectivity seems to be a reasonable requirement of a physical theory.

An additional benefit of mapping to real space-time dimensions is it supports the calculations of fermion-antifermion annihilation in Blaha (2021a).

4.1 NEWQUeST with Four Real Coordinates (Dimensions)

QUeST has four octonion dimensions, which occupy 32 dimensions. We define NEWQUeST to be QUeST with 28 dimensions transformed to serve as the dimensions of new internal symmetries leaving four real space-time dimensions.

We choose to use the 28 dimensions to act as the dimensions of seven U(2) fundamental representations: $U(2)^7$. Each U(2) group serves to map between "normal" and Dark sectors. For each U(2) group joining a pair of blocks: the fermions of the fermion-32 block are individually mapped to the matching fermions of the other block. The dimension-32 internal symmetry groups of the blocks are also so mapped. Fig. 4.1 shows the seven U(2) maps.

Thus NEWQUeST contains eight dimension-32 blocks of symmetries, has four real space-time coordinates (dimensions), and seven additional U(2) groups extracted from the eight dimension-32 blocks. NEWQUeST has 256 dimensions in total. Chapter 8 describes the enhancement of the corresponding UST to NEWUST with $U(2)^7$.

---

[27] In this chapter we take a direct approach. Another procedure would be to reduce octonionic space-time coordinates to real coordinates though some sort of dynamical constructs.

[28] *All new symmetries introduced in this chapter are badly broken except possibly new U(1) symmetries.*

"Normal"   "Dark"

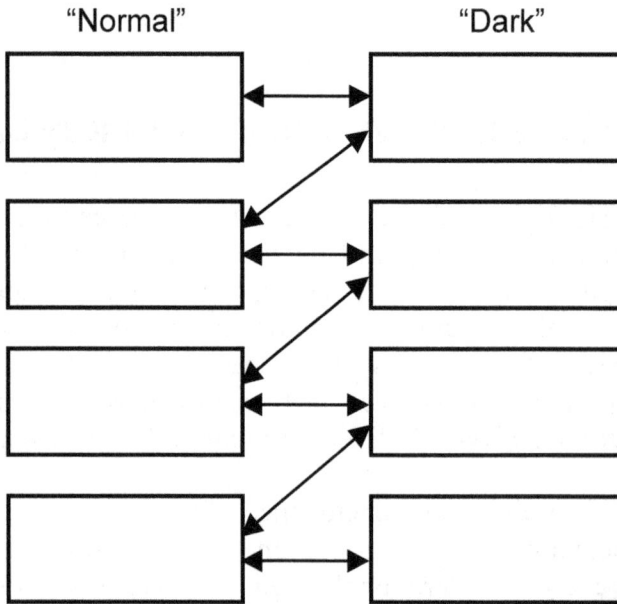

Figure 4.1. Eight dimension-32 atoms arranged in the four layers of the QUeST dimension array. The seven U(2) symmetry groups are extracted from the dimension-32 atoms. They rotate amongst the corresponding fermion-32 blocks. The horizontal lines indicate 1:1 transformations between corresponding fermions of each "Normal" and "Dark" fermion-32 block. The three "angled" lines indicate 1:1 transformations between corresponding fermions of a "Normal" and a "Dark" fermion-32 block in the layer above it. The result is a see-saw pattern.

4.2 NEWUTMOST with Eight Real Coordinates (Dimensions)

UTMOST has eight complex octonion dimensions, which occupy 128 dimensions. We define NEWUTMOST to be UTMOST with 120 dimensions transformed to serve as the dimensions of new internal symmetries leaving eight real space-time dimensions.

UTMOST contains four copies of QUEST, and thus NEWQUeST. Therefore 4* 28 = 112 of the 128 dimensions of UTMOST are to map the four QUeST copies to four NEWQUeST parts of NEWUTMOST. The remaining 16 dimensions serve to give an 8 real dimension space-time plus a U(4) internal symmetry.

Thus NEWUTMOST contains 32 dimension-32 blocks of symmetries, has eight real space-time coordinates (dimensions), and one additional U(4) groups extracted from the 32 dimension-32 blocks. NEWUTMOST has 1024 dimensions in total.

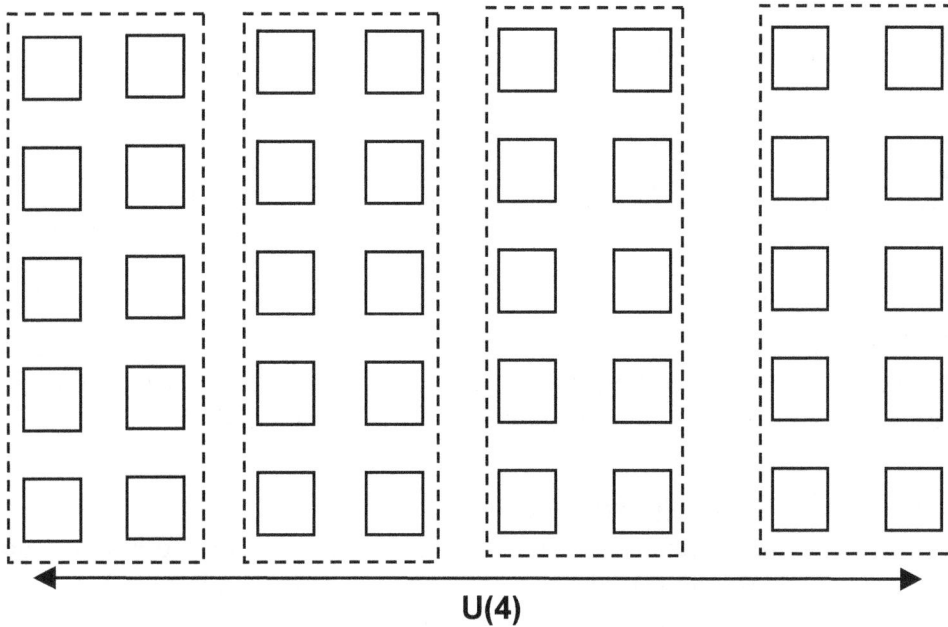

Figure 4.2. NEWUTMOST has four NEWQUeST copies. A U(4) internal symmetry group maps between corresponding fermions in the four copies: fermion by fermion. It also maps between corresponding internal symmetry groups vector bosons.

4.3 NEWMaxiverse with Ten Real Coordinates (Dimensions)

The Maxiverse has ten quaternion octonion dimensions, which occupy 320 dimensions. We define NEWMaxiverse to be the Maxiverse with 310 dimensions transformed to serve as the dimensions of new internal symmetries leaving ten real space-time dimensions.

The Maxiverse contains four copies of UTMOST, and thus NEWUTMOST. The NEWUTMOST's have 4*8 = 32 dimensions allocated to space-time. We reallocate those dimensions to 10 real space-time dimensions plus 22 dimensions for new internal symmetries that we choose to be five additional U(2) groups plus one U(1) group.

Thus the NEWMaxiverse contains 128 dimension-32 blocks of symmetries, has 10 real space-time coordinates (dimensions), and five additional U(2) groups plus one U(1) group extracted from the 128 dimension-32 blocks. NEWMaxiverse has 4,096 dimensions in total.

Fig. 4.3 depicts NEWMaxiverse. In defining the role of the new internal symmetry groups we implement the new rule:

Rule: At most two interactions are needed to connect between fermions in any of four NEWUTMOST parts.

Four NEWUTMOSTs connected by 5 U(2)'s plus a U(1)

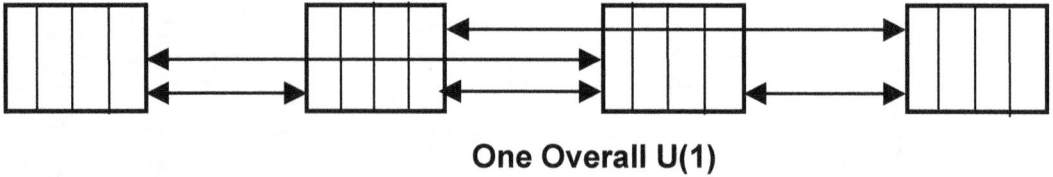

One Overall U(1)

Figure 4.3. NEWMaxiverse consisting of four NEWUTMOST's. Five U(2) symmetries connect corresponding fermions in eac pair of NEWUTMOST's depicted above. A U(1) symmetry exists for all four NEWUTMOSTs.

4.4 Real Space-times for Other Spaces: 3, 2, 1

The other spaces numbered 3, 2, 1, and 0 in Fig. 1.4 are required to have certain space-time dimensions in order to form a sequence of spaces exhibited later in Fig. 6.1.

Space	Required Real Space-time Dim	Array Row/Column Size	Spinor Size
0	18	1024	512 512-spinors
1	16	512	256 256-spinors
2	14	256	128 128-spinors
3	12	128	64 64-spinors
4	10	64	32 32-spinors
5	8	32	16 16-spinors
6	4	16	4 4-spinors
7	2 (built of 4 minispaces)	16	4 4-spinors

Figure 4.4. Table showing the required space-time dimension to have spinors that set the array size of the "next space down." For example space 5 with 8 space-time dimensions has 16 16-spinors. A space 5 fermion-antifermion annihilation generates a space 6 instance whose octonion space has $16^2 = 256$ total dimensions. .

Spaces 0, 1, 2, and 3 can each be viewed as composed of four of the next lower number spaces as we see in Fig. 4.5.[29] We can use the total number of space-time dimensions in the four components of a space to determine the number of dimensions that must be transformed to internal symmetry dimensions (Fig. 4.6) so that the space-time dimensions of a space is that given in Fig. 4.4.

---

[29] For example space 0 has four space 1 parts.

Figure 4.6 shows the number of dimensions transferred to internal symmetries for each space to have real space-time dimensions. The "new" internal symmetry dimensions specify the dimensions of "new" fundamental group representations. We view these "new" groups as supporting communications between the parts of the space as we saw above in the cases of NEWQUeST, NEWUTMOST, and NEWMaxiverse.

As an example we consider the case of the God-Space: space 0 which we suggest supports "communication" interactions between its four component space 1's. See Fig. 4.7.

Space	Space-Time Dimensions	Components
3	12	4 NEWMaxiverses
2	14	4 space 3's
1	16	4 space 2's
0	18	4 space 1's

Figure 4.5. Table of composition of spaces 0, 1, 2, and 3.

Space	Total Space-Time Dimensions Of its Four Components	New Internal Symmetry Dimensions Allocated	Symmetry Before Split
3	40	$40 - 12 = 28$	U(14)
2	14	$48 - 14 = 34$	U(17)
1	16	$56 - 16 = 40$	U(20)
0	18	$64 - 18 = 46$	U(23)

Figure 4.6. Table of new internal symmetry dimensions allocated in spaces 0, 1, 2, and 3. Note the first item in each of the third columns is 4 times the space-time dimension of the previous space's space-time dimension. For example: for space 1 the number 56 equals 4 times 14, which is the space-time dimension of space 2. The fourth column is the symmetry before the split. For example U(14) splits to $U(2)^7$.

4.5 Diagrams of the Connections Generated by the "New" Interactions of Space 0

From Fig. 4.6 we see that 46 dimensions transfer from octonionic space-time dimensions to internal symmetry dimensions for space 0. These dimensions can be viewed as the dimensions of the fundamental representation of U(23) or as a representation of $U(1) \otimes U(2)^3 \otimes U(3)^4$ as we do in Fig. 4.7.

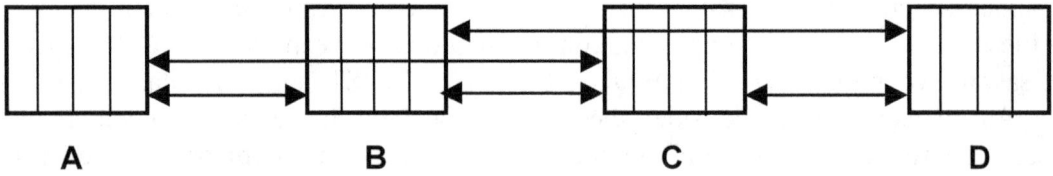

A   B   C   D

Five U(2) groups connecting corresponding fermions in the four space 1 subspaces

PLUS:
One SU(3) for corresponding fermions in A, B, and C
One SU(3) for corresponding fermions in B, C and D
One SU(3) for corresponding fermions in A, B, and D
One SU(3) for corresponding fermions in A, C, and D
One Overall U(1)

Figure 4.7. Possible "new" internal symmetries of NEWGodSpace (space 0) connecting its four space 1 subspaces denoted A, B, C, and D. The U(2) groups connect corresponding fermions in the subspaces. The U(3) groups connect triples of corresponding fermions in three of the subspaces at a time. The U(1) symmetry is global.

5. A Fourth Color – SU(4) in Octonion Space Instances

There have been a number of speculations to the effect that leptons may constitute a fourth color companion of the three known quark colors giving a color SU(4) symmetry group.[30] These considerations are founded upon an *ad hoc* choice of a fundamental symmetry.

While those approaches are of interest, it would be better, in the author's opinion, if one could derive SU(4) from a fundamental principle.

Recently tentative experimental evidence[31] for a fourth color has appeared in a CERN experiment (LHCb collaboration) in rare B-meson decays to electrons and muons where an excess of electrons suggests the possibility (at the 3.1 sigma level) of leptoquark interactions of the form of color SU(4).

Octonion Cosmology offers the possibility of a fundamental justification of SU(4). Fig. 6.1 below displays the group structure of our QUeST and shows repeated appearances of SU(2)⊗U(1), SU(3), and U(1) in its four layers. The combination of SU(3) and U(1) in the figure could indicate an SU(3)⊗U(1) symmetry or could indicate an SU(4) symmetry. QUeST does not specify either choice since QUeST is strictly based on counting dimensions.

The group structure of QUeST, UTMOST and the Maxiverse are based on the structure of the instances of spaces 5 and 6. Chapter 4 of Blaha (2021a) shows that the annihilation of a fermion-antifermion pair leads to a QUeST universe instance with symmetries distributed in 16 dimension blocks. Chapter 8 of Blaha (2021a) shows that the annihilation of a Maxiverse fermion-antifermion pair leads to an UTMOST Megaverse instance with symmetries distributed in 64 dimension blocks.

The 64 and 16 dimension block structuring leads to 64 dimension blocks of layers as illustrated in Fig. 5.1. Sixty-four dimension layers can be subdivided into 16 dimension blocks as shown in Fig. 5.2. This subdivision scenario applies to universes, Megaverses, and the Maxiverse.

The structure of the 16 dimension blocks is:[32]

---

[30] J. C. Pati and A. Salam, Phys. Rev. D10, 275 (1974); C. Quigg and D. Shrock, arXiv: 0901.3958v2 [hep-ph] 2009 and references therein.

[31] CERN seminar 23/03/2012 by K. Petridis and M. Santimaria.

[32] Chapter 3 has an alternate form for 16 dimension blocks. Another alternate form of the 16 dimension blocking that separates "Normal" from "Dark" symmetries is:

> A'. U(4)⊗U(4)
> B'.. U(1)⊗SU(2)⊗U(1)⊗SU(3)⊗(4 dimension real coordinates space-time) *OR*
>   U(1)⊗SU(2)⊗SU(4)⊗(4 dimension real coordinates space-time)

We do not favor this form because it separates space-time symmetries into Normal and Dark parts—an unfavorable form in the author's view.

A. U(4)⊗U(4)

B.. U(1)⊗SU(2)⊗U(1)⊗SU(2)⊗(4 dimension complex coordinates space-time)[33]

C. SU(3)⊗U(1)⊗SU(3)⊗U(1) *OR* SU(4)⊗SU(4).

The content of QUeST universe, UTMOST Megaverses, and the Maxiverse appears in Table 5.1 together with the content of four other octonion spaces.

Space:	6 Complex Octonion Universe Space QUeST	5 Quaternion Octonion Megaverse Space UTMOST	4 Octonion Octonion Maxiverse Space
Dimensions	256	1024	4096
16-Blocks			
Type A	8	32	128 + 6 = 134
Type B	4	16	64 + 3 = 67
Type C	4	16	64 + 3 = 67
64-Blocks	4	16	64 + 3 = 67
Space-Time Dimensions:	4 complex octonion	8 quaternion octonion	10 octornion octonion
Real Dimensions:	64	256	640

Space:	3 Complex Octonion Octonion	2 Quaternion Octonion Octonion Space-timeless Spaces	1 Octonion Octonion Octonion
Dimensions	16,384	65,536	262,144
16-Blocks			
Type A	512	2048	8,192
Type B	256	1024	4,096
Type C	256	1024	4,096
64-Blocks	256	1024	4,096

Space:	0 Superverse Space Complex Octonion Octonion Octonion
Dimensions	1,048,576
16-Blocks	
Type A	32,768
Type B	16,384
Type C	16,384
64-Blocks	16,384

Table 5.1. Tables of the 16-blocks and 64-blocks of the seven octonion space instances.

---

[33] (4 dimension complex coordinates space-time) denotes SL(2, **C**) meaning the SO⁺(1,3) vector representation

5.1 Difference Between SU(4) and SU(3)⊗U(1)

The difference between the choices amounts to the presence or absence of leptonic-quark mixing operators in the group. Only experiment, or further assumptions, can distinguish between the choices:

SU(4)   – Lepton-Quark mixing
SU(3)⊗U(1)   – No Lepton Quark mixing

Theoretic speculations favor SU(4) for "democratic" reasons – certain parity between leptons and quarks as suggested by *Quigg et al* and others. SU(4) would imply the existence of leptoquark (lepton–quark) operators that would cause transitions between leptons and baryons. Such transitions would break baryon number conservation and lepton number conservation. There is no decisive evidence for these transitions at current energies.

On that basis SU(3)⊗U(1) is currently favored.

5.2 SU(4) Fermion Spectrum

The fermion spectrum of a QUeST universe numbers 256 fermions—the same as the number of dimensions. Similarly the number of fermions equals the number of dimensions in the UTMOST Megaverse and Maxiverse instances.

Figs. 5.3 and 5.4 show the close correspondence of the fermion array with the 16 by 16 QUeST array of dimensions. Sixteen dimension blocks of the dimension array correspond directly to 16 fermion blocks. Note that the number of generations is set to four—explaining the number of fermion generations. Also four layers of fermions result thus matching the layers of internal symmetry groups and vindicating the introduction of the four U(4) Layer groups in Fig. 5.1.

The fermion structure of Fig. 5.3 supports SU(4), or SU(3)⊗U(1) in octonion spaces. Thus Octonion Cosmology provides a deep basis for SU(4) and four generations of fermions in four layers. It provides a more satisfying basis for SU(4) than *ad hoc* symmetry-based models.

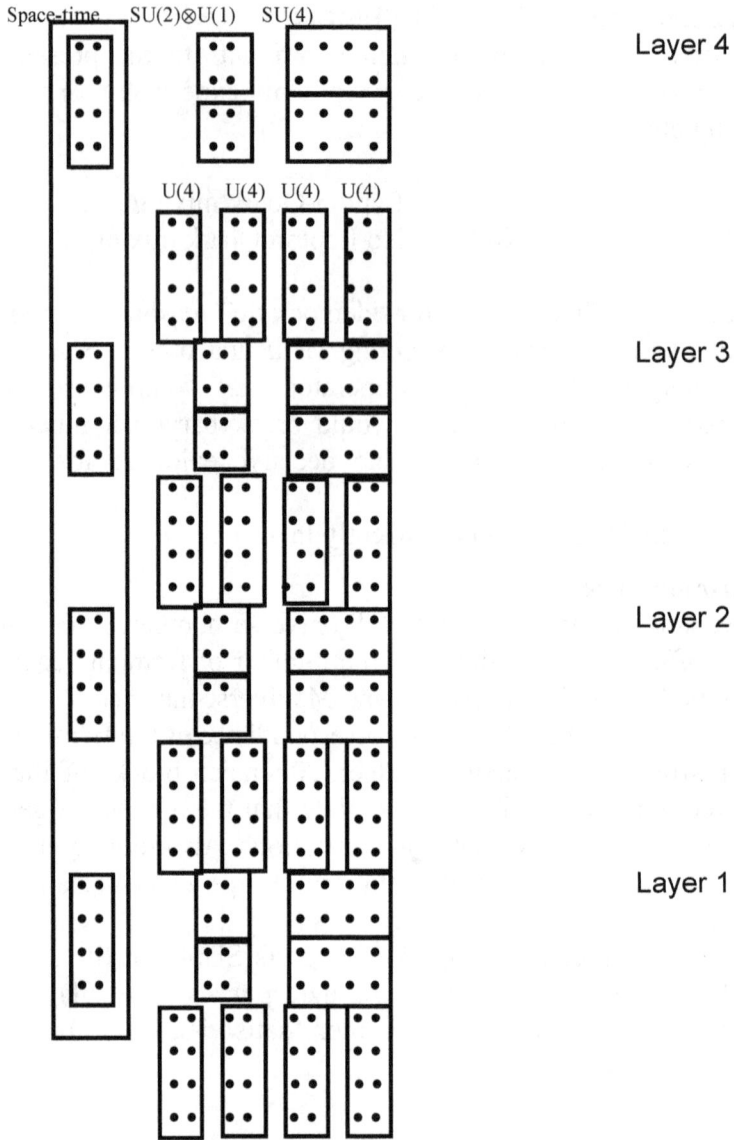

Figure 5.1. The four layers of QUeST internal symmetry groups (and space-time) with SU(4) before breakdown to SU(3)⊗U(1). Note the left column of blocks combine to specify a 4 dimension octonion space-time. Note each layer has 64 dimensions.

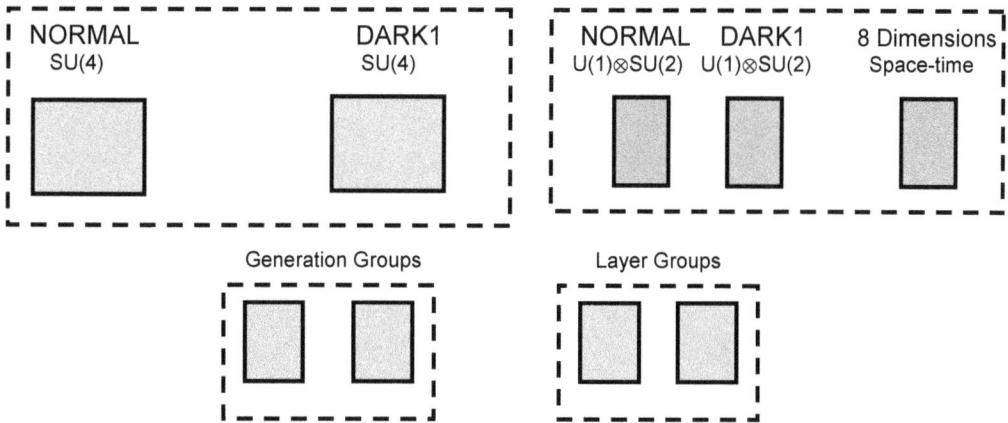

Figure 5.2. A 64 dimension (64-block) block composed of 16 dimension subblocks (encircled by "dotted" lines.) with SU(4) before breakdown to SU(3)⊗U(1). A 64-block forms one layer in QUeST and parts of layers in UTMOST and Maxiverse spaces.

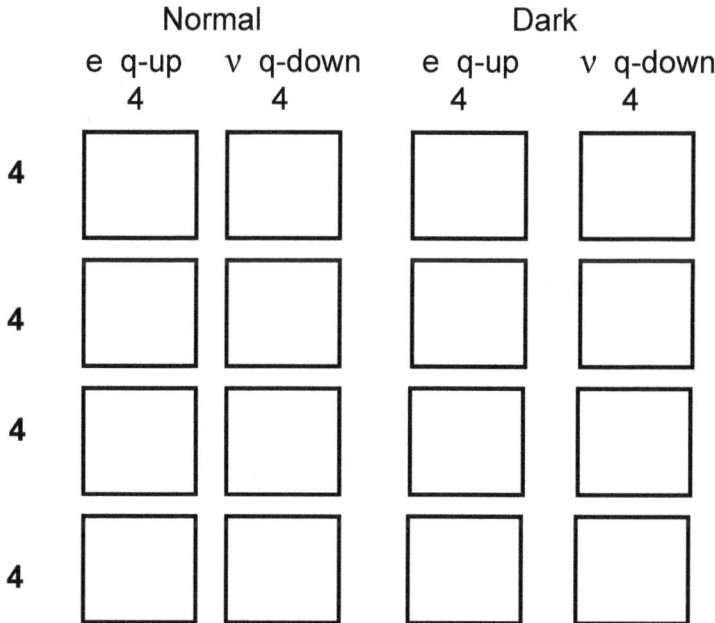

Figure 5.3. Block form of a 16 × 16 QUeST fermion array with each block row corresponding to one layer. Each block contains four generations of fermions. The result is 4 × 4 blocks with SU(4) before breakdown to SU(3)⊗U(1). The label "e q-up" indicates a charged lepton – up-type quark pair, "ν q-down" indicates a neutral lepton – down-type quark pair, and so on.

**OCTONION
DIMENSIONS**

Real   Imaginary

FERMIONS

e   v   up-q   down-q

Layer 1

```
....    ....            .   .   ...    ...
....    ....            .   .   ...    ...
....    ....            .   .   ...    ...
....    ....            .   .   ...    ...
....    ....            .   .   ...    ...
....    ....            .   .   ...    ...
....    ....            .   .   ...    ...
```

Layer 2

```
....    ....            .   .   ...    ...
....    ....            .   .   ...    ...
....    ....            .   .   ...    ...
....    ....            .   .   ...    ...
....    ....            .   .   ...    ...
....    ....            .   .   ...    ...
....    ....            .   .   ...    ...
```

Layer 3

```
....    ....            .   .   ...    ...
....    ....            .   .   ...    ...
....    ....            .   .   ...    ...
....    ....            .   .   ...    ...
....    ....            .   .   ...    ...
....    ....            .   .   ...    ...
```

Layer 4

```
....    ....            .   .   ...    ...
....    ....            .   .   ...    ...
....    ....            .   .   ...    ...
....    ....            .   .   ...    ...
....    ....            .   .   ...    ...
....    ....            .   .   ...    ...
....    ....            .   .   ...    ...
```

Figure 5.4. Fundamental fermions have a 1:1 correspondence with QUeST dimensions. Note the number of dimensions in each row is 8 – the number of dimensions in an octonion. Correspondingly the number of fermions in each row is 8 – a suggestive similarity. Each layer has four normal fermion generations and four Dark fermion generations ("stacked" on each other.) Each dot (pebble) represents a dimension in the left part of the figure and a fermion in the right part.

6. Derivation of the Eight/Ten Octonion Spaces Cosmology

One hopes that ultimately fundamental physics will be reduced to one simple construct. It appears Octonion Cosmology supports that possibility. In chapters 2 and 4 we began considering the interrelation of the octonion spaces.

In this chapter we reduce the eight octonion spaces to one space; space 0 that we call the God-Space since it is the origin of the other spaces through a cascade of fermion-antifermion annihilations that produce instances of the other seven spaces. The seven spaces, expanded to the 10 spaces form of Fig. 1.3, can be viewed as resulting from Type 1 *global splitting by inheritance*. The ten resulting spaces form a subspace of God-Space as shown in Fig. 6.1 below. Thus we see seven/ten octonion spaces are generated from space 0 God-Space.

God-Space generates Octonion Cosmology.

6.1 Features of God-Space   .

We postulate space 0 (God-Space) is the origin of all. It has no beginning or end since it is independent of time. It has a set of internal symmetries that we express in the form of dimension-32 atoms by extrapolating upward from QUeST universe space 6. We found it has 32,768 dimension-32 atoms by eq. 3.3.[34]

The sole instance of God-Space contains a full set of fundamental fermions, vector bosons, and Higgs bosons in analogy with QUeST. We assume that there is only one God-Space instance.

6.2 Generation of Seven/Ten Spaces from God-Space

Since we exist, there must be at least one fermion-antifermion annihilation in God-Space to generate a space 1 instance. The symmetries of space 1 emerge from annihilation by inheritance from the space 0 fermion spinors with space 0 necessarily having 18 space-time dimensions (chapter 4).

Again, since we exist, there must be a cascade of fermion-antifermion annihilations generating a sequence of instances of spaces (with their symmetries) as shown in Fig. 6.2.

Thus we find the eight/ten spectrums of Octonion Cosmology by construction from God-Space – *since we exist,*

For each space one can generate one or more subspace instances. Thus the linear depiction in Fig. 6.2, while correct, does not take account of the possibility of multiple instances of spaces being created. Fig. 6.3 displays a possible hierarchy of instances.

---

[34] See Fig. 5.1 for more a much more detailed list of God-Space symmetries.

6.3 Interconnections of Parts of God-Space and its subspaces

Chaprer 4 shows that the transfer of octonionic space-time dimensions to real Space-time dimensions gives new symmetries, which serve to connect the various parts of each space.

6.4 Hierarchies of Space Instances

Octonion Cosmology supports the creation of a hierarchy of octonion space instances. Each space[35] from space 1 to space 9 may have several instances created by multiple fermion-antifermion annihilations. Hitherto we have pictured a single fermion-antifermion annihilation in a space creating an instance of its subspace. Fig. 6.3 shows a hierarchy containing multiple instances of spaces.

As a result there may be several Megaverses, and each Megaverse can contain multiple universes.

6.5 Analogy to Object-Oriented C++

Hierarchies of instances such as the hierarchy in Fig. 6.3 raise the possibility of an analogy to an Object-Oriented C++ program.[36] We can associate Octonion Cosmology constructs with C++ constructs:

Octonion Cosmology	C++
Space	Class definition
Instance	Object
Subspace	Subclass
Object Hierarchy	Instance Hierarchy
Interactions	Functions
Nested objects	Nested Subspace Instances
Interactions across Sectors	Public Functions
Interactions within a Sector	Private Functions

One sees a similar Space and Class hierarchy. *This correspondence plus other similarities suggests an ultimate view of the Cosmos as a computer program.* Many C++ constructs such as specifications of connections between private and public variables have Octonion Cosmology analogues.

---

[35] We assume space 0 – God-Space has only one instance, from which all lower instances emanate.
[36] See Blaha (1995) for a detailed presentation of Object-Oriented C++ programming.

Superverse – God-Space

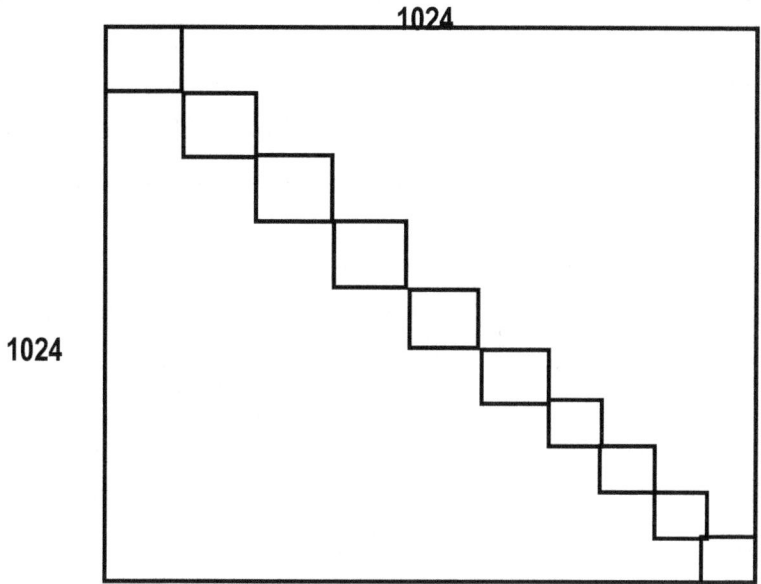

Figure 6.1. The 10 spaces generated from God-Space in block form within its 1024 by 1024 dimension space. The 10 spaces are 512 by 512, 256 by 256, ..., 4 by 4.

God-Space – Space 0: Complex Octonion Octonion Octonion

1,048,576 dimensions	18 Space-time Dimensions	512 512-spinors
$2^{20/2}$ = 1024 rows/columns		

Space 1: Octonion Octonion Octonion

262,144 dimensions	16 Space-time Dimensions	256 256-spinors
$2^{18/2}$ = 512 rows/columns		

Space 2: Quarternion Octonion Octonion

65,536 dimensions	14 Space-time Dimensions	128 128-spinors
$2^{16/2}$ = 256 rows/columns		

Space 3: Complex Octonion Octonion

16,384 dimensions	12 Space-time Dimensions	64 64-spinors
$2^{14/2}$ = 128 rows/columns		

Space 4: Octonion Octonion

4,096 dimensions	10 Space-time Dimensions	32 32-spinors
$2^{12/2}$ = 64 rows/columns		

Space 5: Quaternion Octonion

1,024 dimensions	8 Space-time Dimensions	16 16-spinors
$2^{10/2}$ = 32 rows/columns		

Space 6: Complex Octonion

256 dimensions	4 Space-time Dimensions	4 4-spinors
$2^{8/2}$ = 16 rows/columns		

Space 7: Octonion

64 dimensions	2 Space-time Dimensions	4 4-spinors
$2^{6/2}$ = 8 rows/columns	(Built from four 16 dimension spaces of the 10 space spectrum)	

Figure 6.2. The sequence of fermion-antifermion annihilations and the set of spaces that they generate. The ten space sequence (not shown) has a modified space 7 and adds three more spaces to the sequence as shown in chapter 1..

A COSMOS

SPACE   **INSTANCES**

God-Space – Space 0:

Space 1:

Space 2:

Space 3:

Space 4:

Space 5:

Space 6:   homeverse

Space 7:

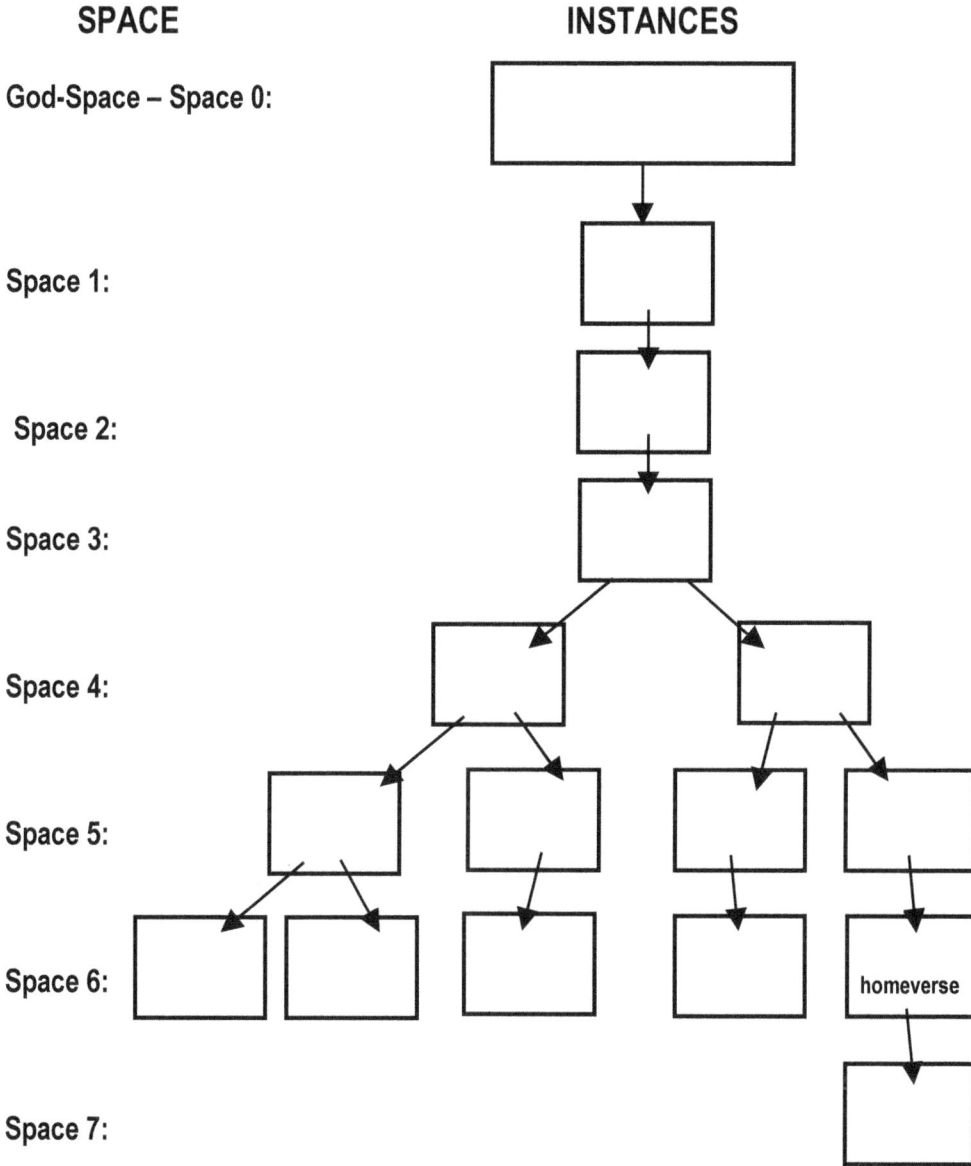

Figure 6.3. A hierarchy of instances leading from the God-Space to "homeverse" – our designation for our universe. The homeverse is shown to contain one space 7 instance for illustration purposes. Space 7 is viewed here as combining spaces 7, 8, 9 and 10, which are part of the 10 space form of the spectrum. The homeverse has one "sibling" and three "cousin" universes. *The entire hierarchy resides in God-Space because the inheritance stems from the God-Space instance as parts of it. Other universes can be "reached" through the God-Space instance if a mode of transportation existed.*

7. A "Universe" in a Photon and Vector Meson Dominance (VDM)

The fermion-antifermion annihilations that we have considered in this book, and in Blaha (2021a), are similarly encountered in e^+e^- annihilation into photons. The difference is e^+e^- annihilation takes place for an on-shell electron and positron.

In a sense a created photon is a microcosm universe that exists for a while until absorbed (or "forever" if not absorbed.) Since it travels at the speed of light it can be viewed internally as frozen in time or everlasting.

We may treat the origin of a photon in a manner similar to the origin of a universe. If we follow the same general procedure as chapter 4 of Blaha (2021a) we obtain a $4 \times 4 = 16$ dimension array from the e^+e^- spinors that is the analogue of the QUeST 256 dimension array. The array, which is a 16 dimension block similar to those in chapter 2, can be represented as composed of fundamental representations of

$$SU(2) \otimes U(1) \otimes SU(2) \otimes U(1) \otimes SU(4)$$

or

$$SU(2) \otimes U(1) \otimes SU(2) \otimes U(1) SU(3) \otimes U(1)$$

Thus we find the produced photons each have an $SU(2) \otimes U(1) \otimes SU(2) \otimes U(1)$ part that dovetails with ElectroWeak vector bosons. They also have an $SU(3) \otimes U(1)$ or SU4) part that dovetails with the Strong interactions, and supports a transition to a ρ vector meson yielding Vector Meson Dominance.

Produced photons are a microcosm of a universe.

8. Augmented UST - NEWUST

The Unified SuperStandard Theory (UST) was shown in Blaha (2020c)[37] to have a four dimension space-time and the internal symmetry

$$[SU(2)\otimes U(1)\otimes SU(3)\otimes SU(2)\otimes U(1)\otimes SU(3)\otimes U(4)^4\otimes U(2)]^4 \qquad (4.4)$$

Remarkably this symmetry was found to be in QUeST as shown in Blaha (2021a) with the change:

$$U(2) \rightarrow U(1)^2$$

giving

$$[SU(2)\otimes U(1)\otimes SU(3)\otimes U(1))\otimes SU(2)\otimes U(1)\otimes SU(3)\otimes U(1))\otimes U(4)^4]^4 \qquad (8.1)$$

or if there is Strong SU(4)

$$[SU(2)\otimes U(1)\otimes SU(4)\otimes SU(2)\otimes U(1)\otimes SU(4)\otimes U(4)^4]^4 \qquad (8.2)$$

QUeST had a four complex octonion dimension space-time.

In chapter 4 we transformed the four complex octonion dimension space-time to a four real dimension space-time plus the new internal symmetries

$$U(2)^7$$

that connect normal and Dark parts of the fermion spectrum. Consequently we now define NEWUST as having a real four dimension space-time and the internal symmetry

$$[SU(2)\otimes U(1)\otimes SU(3)\otimes U(1))\otimes SU(2)\otimes U(1)\otimes SU(3)\otimes U(1))\otimes U(4)^4]^4\otimes U(2)^7 \qquad (8.3)$$

or if there is Strong SU(4)

$$[SU(2)\otimes U(1)\otimes SU(4)\otimes SU(2)\otimes U(1)\otimes SU(4)\otimes U(4)^4]^4\otimes U(2)^7 \qquad (8.4)$$

---

[37] Blaha (2020c) p. 36 eq. 4.4.

9. Possible Evidence for the Existence of a Megaverse

9.1 Theoretical and Experimental Support

Why are we not content with one universe given its enormous size and variety? It appears that there are important theoretical reasons, and some important experimental observations, that suggest that there is more than our universe 'out there.'

In this chapter[38] we will discuss theoretical reasons and experimental suggestions of a larger space—that we call the *Megaverse*—that contains our universe and, most likely, other universes. The existence of a Megaverse resolves several theoretical issues and may address some important astronomical puzzles that have appeared in recent years.

The theoretical issues, which have been subjects of discussion for many years, are:

1. The need for a 'clock' to measure 'time' knowing that it is to some extent relative and local.
2. The need for a 'quantum observer' to complete the understanding of quantum gravity as described by the Wheeler-DeWitt equation and in other efforts to develop a quantum gravity.
3. The need for other universes to provide theoretical measuring platforms for quantities beyond the charge and mass of the universe. We think here of the other quantum numbers of particles and particle number operators such as Baryon number.
4. The need for an ultimate source of mass and inertia in our universe.

In Blaha (2015a) and earlier books we have suggested that there are weighty reasons to believe that other universes exist.[39] The existence of other universes is a solution to these problems.

These problems have a source in Quantum Gravity and the interpretation of the Wheeler-DeWitt equation in particular. We now consider the issues raised above.

9.1.1 Universe Clocks

Asynchronous Logic provides the equivalent of a clock for the synchronization of processes within large electrical systems such as VLSI chips. Similarly there is a

---

[38] Most of this chapter appears in Blaha (2015a) and in earlier books by the author.
[39] In Blaha (2013a), before the Higgs particle was discovered at CERN we suggested an alternate mechanism was possible if a sister universe existed (making the existence of other universes a reasonable possibility. The Higgs discovery makes the sister universe mechanism unlikely.

need for a universal clock for our universe. As DeWitt[40] points out in his studies of quantum gravity,

"'The variables ... [of the quantized Friedmann model] because of their lack of hermiticity, are not rigorously observable and hence cannot yield a measure of proper time which is valid under all circumstances. ... . It is for this reason that we may say that "time" is only a phenomenological concept ... If the principle of general covariance is truly valid then the quantum mechanics of everyday usage with its dependence on the Schrödinger equations ... is only a phenomenological theory. For the only "time" which a covariant theory can admit is an intrinsic time defined by the contents of the universe itself. Any intrinsically defined time is necessarily non-Hermitean, which is equivalent to saying that there exists no clock, whether geometrical or material, which can yield a measure of time which is operationally valid under *all* circumstances, and hence there exists no operational method for determining the Schrödinger state function with arbitrarily high precision."

The lack of a clock within our universe invalidates quantum mechanics in principle and Quantum Gravity in particular. DeWitt concludes, "Thus [quantum gravity] will say nothing about time unless a clock to measure time is provided."

Unruh[41] also has an issue with the source of time:

"One of the key problems is that of time. We see and experience the world in terms of time. We see things grow, develop, and change. However, time does not enter into the Euclidean formulation of quantum gravity directly. In the usual Hamiltonian formulation, the Hamiltonian for quantum gravity is made up of densities which are the generators, not only of spatial coordinate transformations, but also of temporal coordinate transformations. The content of four of Einstein's equations is that some generators are zero. Thus all wave functions are invariant under all spatial and all temporal coordinate transformations. There is nothing in the wave function or the amplitudes which refers to the coordinate t, or the corresponding points of the manifold in any way. How then do we recover the indubitable and ubiquitous experience we have of time? The standard answer is that our experience of time is actually an experience of different correlations between physical quantities in the world. Time is replaced by the readings of clocks. I know that time has changed, not through any direct experience with time, but because the hands of my watch have changed.

Although the implementation of this idea is actually extremely difficult in practice, and although I personally believe that one should formulate one's quantum theory of gravity so as to contain time explicitly, let us nevertheless pursue the consequences of this idea of time as defined internally, as the "reading" of a dynamic variable. For an observer inside the theory, his "time" is not the coordinate t. Rather his

---

[40] DeWitt, B. S., Phys. Rev. **160**, 1113 (1987).
[41] Unruh, W. G., Phys. Rev. D **40**, 1053 (1989).

time is some one of the given dynamic variables of the theory: y or P. Thus although the coupling to the baby universes via the effective action S is independent of the coordinates t or x, that does not mean that the observer inside the theory will experience the interactions as being independent of time. For him and/or her, time is one of the dynamic variables and so it can depend on the various dynamic variables of the theory, even if it does not depend on the time coordinate t. In general one would expect the observer to see what looks to him like a time-dependent interaction with the baby universes. At one time, some one of the baby universes may couple strongly to the large universe, while at some other time, another of the baby universes will couple more strongly."

In Blaha (2015a) and earlier books, we suggested the existence of other universes provides a 'clock' in principle for our universe. And being universes, these other universes are excellent clocks. DeWitt points out,

"Because every clock has a "one-sided" energy spectrum, its ultimate accuracy must necessarily be inversely proportional to its rest mass. When the whole universe is cast in the role of a clock, the concept of time can of course be made fantastically accurate (at least in principle) ... "

Setting a mass scale using other universes, also sets[42] a time scale and resolves the issue of a clock for our universe. *In principle the existence of other universes validates the role of time in the Copenhagen interpretation of Quantum Mechanics.*

9.1.2 Quantum Observer

Attempts to create a quantum gravity theory have to confront the need for an *Observer* in any quantum theory within the context of the Copenhagen interpretation. DeWitt points out,

"The Copenhagen view depends on the assumed a priori existence of a classical level to which all questions of observation may ultimately be referred. Here, however, the whole universe is the object of inspection; there is no classical vantage point, and hence the interpretation question must be re-argued from the beginning. While we do not wish to stress this point unduly, since, after all, the Friedmann model ignores the vast complexities of the real universe, it is nevertheless clear that the quantum theory of space-time must ultimately force a deviation from the traditional Copenhagen doctrine." And Unruh states

"One of the key features in the interpretation of such transition amplitudes, or wave functions, is the idea that we, as observers are also a part of the Universe as a whole. We, as physical observers, must be describable from within the theory and not as

---

[42] For example the Planck time value is set by the Planck mass.

observers external to the theory as in usual quantum mechanics. In usual quantum mechanics, the interpretation is usually given in terms of observers that are outside of the theory. There one makes a split, with the quantum world at one side of the split, and the observer on the other. von Neumann argued that the predictions of quantum mechanics, at least under certain assumptions, are independent of the exact location of that split, but Bohr argued adamantly for the necessity of such a split (classical observers and quantum world). *There is a great difficulty in setting up such a split for physical observers contained within and influenced by a quantum universe,* [italics added] and for the Universe as a whole, especially including gravity, one cannot argue that the predictions will be independent of where one puts the split. Since all energies interact gravitationally, and our observations are surely energetic phenomenon, the treatment of the energetics of observation as classical would lead to different predictions than if they were treated quantum mechanically. One is therefore forced to devise an interpretation of quantum mechanics in which the observer is part of the quantum system, rather than outside the quantum system.

This means that the interpretation of these transition amplitudes becomes somewhat non-intuitive. One must ask what the system looks like from within, from the viewpoint of an observer who is part of that world, rather than being able to interpret them directly in terms of probabilities for observations made by an external observer."

While the O*bserver* question is addressed by a number of authors, the proposed answers are not entirely convincing. *The existence of other universes provides macroscopic Quantum Observers for our universe.* And our universe provides a macroscopic quantum observer for other universes. Thus the quantum observer issue is resolved.

These considerations lead us to view the existence of other universes as a critical solution to the above problems.

9.1.3 The Higgs Mechanism is Explainable by Extra Dimensions

The Higgs Mechanism 'explains' (generates) fermion and boson masses. However the Higgs potential contains a quadratic term with a constant with the dimensions of [mass]. In a sense the Higgs Mechanism trades one mass for another. From where do the Higgs potentials' masses come?

A further explanation is needed is to determine the origin of the "dimensionful" mass terms in the Higgs' particle equations themselves. At present little if any thought has been given to the origin of these terms. We suggested that, excluding a *deus ex machina* source, the only known way to generate these mass terms in the Higgs' equations is through the separation of equations technique of differential equations. This technique requires additional parameters which can only be the coordinates of *extra unknown dimensions*. The best example of the generation of mass terms appears in the Schwarzschild solution of General Relativity where a separation constant, often denoted M, appears that has the dimension of [mass].

Thus extra space-time dimensions would resolve the origin of Higgs potentials' masses. Given extra dimensions it is reasonable to expect that these extra dimensions contain universes. Thus the Megaverse!

9.1.4 Possible Accretion of Megaverse Matter to Fuel Expansion of Our Universe

If matter is distributed outside of universes in the Megaverse, and if this matter can be accreted to universes by gravitational attraction, then the apparent increasing expansion of our universe may be due to this accretion. In chapter 14 of Blaha (2017c) we presented a model in which this possibility is realized. If true, then we would have tangible evidence of the residence of our universe in the Megaverse.

9.1.5 Asynchronous Logic is a Requirement of Universes

By establishing Asynchronous Logic principles[43] as the basis for the existence of universes and for setting the number of dimensions in each universe – four; and basis of fermion particles - qubes – we have found deeper principles of organization for the foundations of physics. The principles built on this foundation serve to enable the coordination of complex physical processes.

Usually we look at particle processes primarily from a space-time perspective: particles collide and produce new particles. We primarily think of the incoming and outgoing particles in a collision. However, considering the set of fundamental particles – and the particle transforming interactions in themselves – neglecting space-time and momentum considerations – leads us to view particles as constituting an alphabet and their interactions as a type of computer grammar.[44] Then the Asynchronicity Principles enable us to bring in space-time in a way that gives us the maximum complexity with the most minimal assumptions. As Leibniz[45] points out our universe has maximal complexity with minimal assumptions.

9.1.6 The Meaning of Total Quantities of a Universe

The 'external' properties of a universe are normally questioned—for the simple reason that it is assumed that there is no 'outside' of our universe. For example, Misner (1973) asserts:[46]

'There is no such thing as "the energy (or angular momentum, or charge) of a closed universe," according to general relativity, and this for a simple reason. To weigh something one needs a platform on which to stand to do the weighing.'

Misner et al presumes no such platform exists. If there is but one closed universe as most currently believe then one cannot measure any totals of a closed universe (which

---

[43] The basis of this section is described in detail in Blaha (2015a). That book places Physics within a logical framework that is a possible deeper ground for fundamental Physics theory.

[44] This conceptual approach was first described in Blaha (1998) who went on to characterize our universe as one enormous word evolving in time.

[45] See Rescher (1967).

[46] Pp. 457 - 458.

ours may be to be). Yet if we take a more general view that our universe is only one of many then it becomes possible to measure total mass, charge, angular momentum, baryon number, and many other quantities of interest. Indeed, the existence of other universes (within the encompassing Megaverse) opens the door to an understanding of time, mass, energy, and all the other quantities necessary to develop a dynamical theory of universes.

Later we will also see that one can then treat universes as 'particles', and develop 'universe dynamics', which might explain knotty problems such as the Big Bang and its precursor (if any). We will do this in subsequent chapters after first considering the possible structure of universes in general in the Megaverse.

9.2 Possible Experimental Evidence for the Megaverse

At first glance it would seem impossible to produce evidence for the existence of other universes. However there are subtle means by which we can 'sense' experimentally 'nearby' universes should they exist. The mechanism would appear to be gravitational effects exerted on objects within our universe by unseen objects of enormous mass. Currently there appears to be three experimental suggestions of the existence of 'nearby' universes and one theoretical argument based on an influx of mass-energy from the Megaverse that may cause the expansion of our universe.

9.2.1 Great Attractors

One potential support is the discovery of the Great Attractor (at the center of the Laniakea Galaxy Supercluster), and the more massive Shapley Attractor (centered in the Shapley Supercluster)[47]. These attractors contain massive numbers of galaxies and are drawing galaxies over a distance of millions of light years towards them.

If another universe(s) is 'near' our universe it could act as a 'gravitational magnet' and draw galaxies within our universe towards it to form one or more superclusters which could then act as attractors. Thus attractors might indirectly reveal the presence of other nearby universes—contrary to the expected large scale uniformity of the universe. The only other apparent source of superclusters is chance. Chance seems an unsatisfactory possibility in the present case.

9.2.2 Bright Bumps in Universe Suggesting Collision with Another Universe

A recent study[48] of the residual brightness of parts of the accessible universe found that bright patches appeared if a model of the CMB (Cosmic Microwave Background) with gases, stars and dust was 'subtracted' from the PLANCK map of the entire sky. After the subtraction one would expect only noise spread throughout the sky. However, bright patches were seen in a certain range of frequencies. These anomalies are thought to be a result of our universe colliding with another object – presumably another universe in the Megaverse.

---

[47] Tully, R. Brent; Courtois, Helene; Hoffman, Yehuda; Pomarède, Daniel, "The Laniakea Supercluster of galaxies". Nature (4 September 2014). 513 (7516): 71–73; arXiv:1409.0880.
[48] Ranga-Ram Chary, arXiv.org:/1510.00126 (2015).

9.2.3 Cold Spot in Universe Suggesting Collision with Another Universe

Another recent study[49] of a huge cold region of the universe spanning billions of light years revealed that this region is not a relatively empty region but rather is similar to in its distribution of galaxies to the rest of the universe. Previous the Cold Spot (an area where cosmic microwave background radiation – the leftover Big Bang radiation is weak – making it significantly colder (0.00015C colder) than the average temperature of the universe.)

An analysis of 7,000 galaxy redshifts using new high-resolution data has now shown that the Cold Spot is similar to the rest of the universe. The Durham University group suggested that the Cold Spot might have been caused by a collision between our universe and another Universe. They further suggested that there is only a 1 in 50 chance that it could explain by standard cosmology. could produce this feature

Thus we have another important piece of circumstantial evidence in favor of other universes and thus the Megaverse.

9.2.4 Megaverse Energy-Matter Infusion into Our Universe

In chapter 14 of Blaha (2017c) we presented a model for an influx of mass-energy from the Megaverse to support the Bond-Gold-Hoyle-Narlikar Steady State Cosmology, which was originally based on the 'continuous creation of mass-energy' by Hoyle and Narlikar. This model explains why the value of Ω makes the universe close to flat. If this model is correct then we would have concrete support for a Megaverse with a low mass-energy density leaking mass-energy into our universe. *More generally, it suggests that universes are surfaces of high mass-energy density in a Megaverse of low mass-energy density – with a ratio of mass-energy densities of the other of 10^{30}.*

9.2.5 Conclusion

We conclude that data is beginning to emerge favoring multiple universes and a physical Megaverse in support of the theoretical justifications presented earlier.

---

[49] T. Shanks et al, Durham University (Australia), Monthly Notices of the Royal Astronomical Society, 2016 .

10.Growth of Octonion Space Instances from a Big Bang

Universes and Megaverses grow from annihilation events in a Megaverse and a Maxiverse respectively. In this chapter we will consider the growth of universes and Megaverses (and possibly other space instances) from the initial annihilation based on the assumption of a generalization of coordinates in the created instance to Two Tier quantum coordinates that the author developed many years ago. Quantum coordinates "smear" coordinates to prevent divergences in quantum field theory calculations,[50] and, in the present case, to prevent a catastrophe at the "Big Bang" point of the instance.[51]

Later in this chapter we will consider a generalization of the formula for the universe's scale factor and Hubble Parameter to accommodate the Big Bang period as well as more recent times. This formula can be based on a universe/Megaverse model with a "vacuum polarization" that is analogous to that in Blaha's exact calculation of the Fine Structure Constant, α, based on vacuum polarization in the Johnson-Baker-Willey formulation of Quantum Electrodynamics. *It is a model based on a close analogy (conceptually and numerically) with the Quantum Electrodynamics vacuum polarization of a particle. We believe an octonion space instance is a type of particle.*

Our interim Big Bang Model was based on a quantum coordinate that is defined using a massless vector field denoted $Y^{\mu}(y)$:

$$X^{\mu}(y) = y^{\mu} + i\, Y^{\mu}(y)/M_c^{2} \qquad (10.1)$$

where the y^{μ} coordinates are ordinary "point" coordinates.

In Blaha (2019e) we calculated the expansion of a universe (or a Megaverse) from the "Big Bang" point and showed the "quantum pressure" due to the Planck Black Body energy distribution of the $Y^{\mu}(y)$ fields prevented an infinite collapse at t = 0. We then matched the scale factor and Hubble Parameter of our Big Bang model with the empirically found values that we found it described the universe subsequently. Here we extend our empirical model to the Big Bang of this, and other octonion space, instances.

10.1 Growth Scenario

Growth begins with the transition of the fermion-antifermion pair to a universe (or Megaverse) particle. The fermion antifermion pair use Two-Tier quantum coordinates and the created particle's coordinates are TwoTier coordinates in its interior.

---

[50] Blaha (2002), and (2005a).
[51] Blaha (2004) and (2019e).

The particle contains an extremely large internal energy distribution that primarily resides in the Planck Black Body energy distribution of the $Y^\mu(y)$ fields. The energy distribution initially stabilizes the particle to a finite size preventing a collapse at $t = 0$.

The particle then expands in a Big Bang generating the growth of the particle (universe or Megaverse).

10.2 The Blaha (2019a) Calculation of Growth

The calculation of the growth of the universe in the interim model[52] showed an enormous growth as expected.

The radius of the universe is non-zero as expected due to the use of Two-Tier Quantum coordinates. At $t = 0$

$$r = 4.278 \times 10^{-65} \text{ cm} \qquad (10.2)$$

and at time $t_c = 1.26 \times 10^{-165}$ s

$$r = 8.5 \times 10^{-65} \text{ cm} \qquad (10.3)$$

with an effective doubling of the radius.

The scale factor $a(t)$ at time $t = 0$ is

$$a(0) \cong 3.19 \times 10^{-93} \qquad (10.4a)$$

and at $t_c = 1.26 \times 10^{-165}$ s is

$$a(t_c) \cong 1.632 \times 10^{-92} \qquad (10.4b)$$

The Hubble Parameter at $t = 0$ and $\check{r} = 1$, the scaled radius at the edge of the Big Bang, is

$$H_{BBRW}(0, 1) \cong 1.79 \times 10^{218} \text{ km s}^{-1} \text{ Mpc}^{-1} \qquad (13.4.7)$$

(compared to the current Hubble constant of $100h$ km s^{-1} Mpc^{-1}.

At $t = 0$ and $\check{r} = 0$,

$$H_{BBRW}(0, 0) \cong 2H_{BBRW}(0, 1) \qquad (13.4.8)$$

At the "edge" of the universe at time $t_c = 1.26 \times 10^{-165}$ s and $\check{r} = 1$ the Hubble Parameter is[53]

---

[52] Sections 10.2-10.4 quotes Blaha (2019c)'s text and equations using its equation numbers.
[53] Using eq. 13.3.2.15. of Blaha (2019e)

$$H_{BBRW}(t_c, 1) \cong 1.14 \times 10^{126} \text{ km s}^{-1} \text{ Mpc}^{-1} \qquad (13.4.9)$$

At at time $t_c = 1.26 \times 10^{-165}$ s and $\check{r} = 0$, the "center" of the Big Bang has

$$H_{BBRW}(t_c, 0) \cong 8.95 \times 10^{217} \text{ km s}^{-1} \text{ Mpc}^{-1} \qquad (13.4.10)$$

using eq. 13.3.2.14. *The "center" is more rapidly expanding by a factor of the order of* 10^{93}, *which is suggestive of a developing, much less dense region.*

10.2.1 Creation of a Void in the Metastable State due to a Radially Varying Hubble Constant

Sections 13.6.1 and 13.6.2 show that at $t = 0$ the Hubble Parameter is a factor of 2 greater at $\check{r} = 0$ than at the edge of the universe ($\check{r} = 1$), and at $t = t_c$ (the end of the metastable state life) the Hubble Parameter is a factor of the order of 10^{93} greater at $\check{r} = 0$ than at the edge of the universe ($\check{r} = 1$). The vast disparity in Hubble Parameter values at the center compared to the edge of the universe suggests the center will rapidly go to lower density creating a *void. Thus one can expect a "Hubble bubble" generated during the metastable state.*

Today we see numerous voids in the universe.[54] The largest is the spherical KBC supervoid (containing the Milky Way) with a 2 billion light year diameter. Another supervoid is the "Giant void" (Canes Venatici) with a diameter of 1.3 light years. There is also the WMAP "cold spot" void with a diameter of 120 Megaparsecs.

The Hubble Parameter within this void is larger due to the attraction of mass-energy external to the void just as we see above when comparing the center to the edge of the Metastable state.

The metastate "void" appears to be a precursor to the later voids seen now. Voids may also be the result of the fluctuation in universe size around the time of the Big Dip.

10.3 Original Phenomenological Model to the Hubble Parameter

Another calculation of a model for the change of the Hubble Parameter over time used the experimentally determined values

$$H(t_c) \equiv H(380,000 \text{ yr}) = 67.8 \qquad (10.5)$$
$$H(t_{now}) = 73.24$$

These values approximate the set of known experimentally determined values. The phenomenological model was based on the assumptions:

---

[54] Zehavi *et al*, Astrophysical Journal **503**, 483 (1998); There are differing results on the Hubble Bubble question such as in Moss *et al*, Phys. Rev. **D83**, 103515 (2011) and references therein.

Assumptions:

"The radiation-dominated and the matter-dominated scale factors a(t) both are power laws in time as seen earlier in Blaha (2019c). We therefore will assume that the true a(t) has a power law form:

$$a(t) = (t/t_{now})^{g + ht} \qquad (22.1)$$

where g and h are constants, and $t_{now} = 4.35 \times 10^{17}$ s. (The constant h is *not* the Hubble parameter.) There is an "ht" term in the exponent based on the rise in H(t) noted above in the experimental data.

The bases of this choice are:

1. Power law behavior (in part) as in the radiation and matter dominated approximations seen earlier.

2. The known shape of H(t) at early times, and at present. Eq. 22.1 is consistent with the shape of H(t).

3. The simplicity of the model. Two values of H(t) set the constants g and h.

4. Faster than exponential future growth as $t \to \infty$.

$$a(t) = \exp[(g + ht)\ln(t/t_{now})] \sim e^{ht \, \ln(t)}$$

5. *The small time behavior of a(t) can be derived in a particle model of a universe as shown in detail later and in Appendix 10-A as well as Blaha (2019c).*

The Hubble Constant implied by eq. 22.1 is

$$H(t) = (da/dt)/a = g/t + h(1 + \ln(t/t_{now})) \qquad (22.2)$$

If we set the value of H(t) at two values of time, then g and h are determined. Based on the above discussion we use the experimental data:

$$H(380,000 \text{ yr}) = 67.8 \qquad (22.3)$$
$$H(t_{now}) = 73.24 \qquad (22.4)$$

Eqs. 22.2 and 22.3 imply

$$h = (t_c H(t_c) - t_{now}H(t_{now}))[t_c - t_{now} + t_c \ln((t_c/t_{now})]^{-1} \qquad (22.5)$$
$$g = (H(t_{now}) - h) t_{now}$$

Substituting the parameter values of eq. 22.3 we obtain[55]

$$h = 2.25983 \times 10^{-18} \qquad (22.6)$$

$$g = 0.000282377 = 2.82377 \times 10^{-4}$$

Since

$$h \cong 1/t_{now}$$

an alternate possible form for a(t) is

$$a(t) = (t/t_{now})^{g + t/t_{now}} \qquad (22.1a)$$

We found this form to be suggestive but not consistent with the current values of the Hubble Parameter.

If one wanted the value of h to be equal to 1//t$_{now}$ then the present value of the Hubble Parameter would have to be 74.47 (if the Hubble value at 380,000 years is 67.8) – a value within the range of experimental values."

10.3.1 Original Model a(t) at Small Times (the Big Bang Period)

The original model for a(t), eq. 22.1, seems to work well for large times greater than 380,000 years. However at very small times such as the Big Bang period a(t) must be modified since

$$a(0) = 0 \qquad \text{BAD} \qquad (10.6)$$

at t = 0 resulting in a divergence. H(t) is also divergent at t = 0.

10.4 A New Phenomenological Model Extending to Big Bang Period

If we examine a(t) for small times we find eq. 22.1 above we find

$$a(t) \sim (t/t_{now})^{g} \qquad (10.7)$$

where the constant g can be determined by a vacuum polarization calculation as seen in Blaha (2019c) and Appendix 10-A below. Furthermore the numeric value of g is also determined in that calculation. See appendix 10-A.

A study of the Johnson-Baker-Willey model by Adler[56] suggested that the vacuum polarization summed to all orders might have an essential singularity at α perhaps of the form

$$\exp[-1/(\alpha - 0.0072973525693)] \qquad (10.8)$$

where α has the known 0.0072973525693.[57]

---

[55] **The calculation of g, in particular, is delicate since it contains small differences between large quantities.**
[56] S. Adler, Phys. Rev. **D5**, 3021 (1972).

Motivated by this suggestion, and the successful understanding of eq. 10.7 from vacuum polarization, we propose a(t) has an additional factor that governs its behavior in the Big Bang period:[58]

$$a(t) \cong [(t + t_0)/t_{now}]^g [(t + t_0)/t_{now}]^{gd/(t + t_0)}$$
$$= [(t + t_0)/t_{now}]^{g [1 + d/(t + t_0)] + ht} \qquad (10.9)$$

where t_0 is a base time value, and d is a constant. The new a(t) has an essential singularity at the unphysical $t = - t_0$. Since t_0 will be seen later to be extremely small it is possible to view the state at $t = - t_0$ as a precursor state of a fermion-antifermion annihilation as we did earlier in Octonion Cosmology.

The resulting Hubble Parameter is

$$H(t) = [g + ht + gd/(t + t_0)]/(t + t_0) + [h - gd/(t + t_0)^2] \ln[(t + t_0)/t_{now}] \qquad (10.10)$$

At t = 0

$$a(0) = (t_0/t_{now})^{g[1 + d/t_0]} \qquad (10.11)$$

$$H(0) = [g + gd/t_0]/t_0 + [h - gd/t_0^2] \ln(t_0/t_{now}) \qquad (10.12)$$

Note a(0) and H(0) are not zero, thus avoiding a Big Bang catastrophe.

For large t

$$a(t_{now}) = 1 \qquad \text{Normalization} \qquad (10.13)$$

$$a(t) \cong (t/t_{now})^{g + ht} \qquad (10.14)$$

as in eq. 22.1 assuming d and t_0 are small with $d/t \ll 1$ and $t \gg t_0$. For large t

$$H(t) = g/t + h(1 + h \ln(t/t_{now})) \qquad (10.15)$$

as in eq. 22.2. Eq. 10.15 follows if d = 0 is substituted in eq. 10.10.

10.5 A Universe Quantum Vacuum Polarization Basis for the New Model

We now proceed to determine the constants d and t_0 since g and h are determined by the large time behavior of H(t). The equations fixing d and t_0 are

$$a(0) \cong (t_0/t_{now})^{g[1 + d/t_0]} = a_{BBRW}(0) \cong 3.19 \times 10^{-93} \qquad (10.16)$$

$$H(0) = [g + gd/t_0]/t_0 + [h - gd/t_0^2] \ln(t_0/t_{now}) = H_{BBRW}(0,0) \cong 3.58 \times 10^{218} \text{ km s}^{-1} \text{ Mpc}^{-1}$$
$$= 1.1 \times 10^{199} \text{ s}^{-1} \qquad (10.17)$$

---

[57] Calculated to the exact known experimental value in Blaha (2019f).
[58] We choose $t + t_0$ rather than $t - t_0$ to avoid an essential singularity at positive t.

For small $t_0 < 10^{-100}$ s we find $H(0)$ is well approximated by

$$H(0) \cong -(gd/t_0^2)\ln(t_0/t_{now}) \tag{10.18}$$

up to a few per cent. Taking the logarithm of $a(0)$ we see

$$\ln(a(0))/H(0) = -t_0(t_0/d + 1) \tag{10.19}$$

From

$$\ln(a(0)) = g(1 + d/t_0)\ln(t_0/t_{now}) \tag{10.20}$$

we see

$$d = t_0 [\ln(a(0))/\ln(t_0/t_{now}) - g]/g \tag{10.21}$$

As a result

$$\ln(a(0))/H(0) = -t_0(1 + 1/((1/g)(\ln a(0)/\ln(t_0/t_{NOW})) - 1)) \tag{10.22}$$

which fixes t_0 to be

$$t_0 = 1.936 \times 10^{-197} \text{ s} \tag{10.23}$$

and

$$d = 2.956 \times 10^{-194} \text{ s} \tag{10.24}$$

by eq. 10.21. Eq. 10.23 numerically is

$$\ln(a(0))/H(0) \cong -t_0 \tag{10.25}$$

to good approximation, which implies

$$a(0) = \exp(-t_0 H(0)) \tag{10.26}$$

a form similar to quantum mechanical expressions. In the Big Bang region ($t < t^{-200}$ s) $a(t)$ and $H(t)$ are constant by Fig. 10.2 so

$$a(t) = \exp[-(t + t_0)H(t)] \tag{10.27}$$

for $t < t^{-200}$ s since $t_0 = 1.936 \times 10^{-197}$ s dominates in eq. 10.27.

10.6 A Possible of the Hubble Parameter as a Hamiltonian in the Big Bang Region

The Hubble Parameter is viewed as a time dependent variable. Eq. 10.27 raises the possibility that it might be treated as a time dependent Hamiltonian, if appropriately formulated, with eq. 10.27 generalized to

$$a(t) = \exp[-(t + t_0)H(t)] \tag{10.28}$$

for "all" time. Then a(t) becomes a kind of wave function, while retaining its role as the scale factor of the universe. This possibility, which is evocative of our QED-like Model, remains to be explored in detail.

10.7 Consequences of the New Blaha Model

Fig. 10.1 shows the general form of our model for the a(t) scale factor from t = 0 to the present. Figs. 10.2-10.7 show a plot of the Hubble Parameter of eq. 10.10 from t = 0 to the present.

The model has an impressive set of features:

1. The scale factor becomes very small, but non-zero, as t → 0, the Big Bang point. The scale factor a(t) and H(t) are constant in the Big Bang region from t = 0 to 10^{-200} s. The essential singularity part with the parameter d greatly affects a and H in the Big Bang region.

2. H becomes very large as t → 0.

3. The large time features of a and H are consistent with known data. See Appendix 10-A.

4. From t =o to t = 10^{-200} s a(t) and H(t) are flat. This region is the Big Bang region.

5. The constants g and h are determined from H in the large t region. The parameters d and t_0 are determined by a(t), H and g in the Big Bang region.

6. The essential singularity at t = $-t_0$ *may* represent the point of fermion-antifermion annihilation to create the universe or Megaverse instance.

Thus we have a satisfactory set of parameters for the new scale factor model extending to the Big Bang period.[59] The additional factor with an essential singularity in time yields a satisfactory model for t = 0 to the present.

10.7.1 A Big Dip in H(t)

Fig. 10.5 shows a Big Dip that we have seen before in earlier work. This feature is to be expected since H(t) declines from a large positive value at small times and rises at large times near the present. *A minimum must be present by simple algebra.*

The Big Dip events took place at:

Big Dip minimum (H ≅ -410) at t = t = 4.1199×10^{14} sec.

---

[59] The values of a(0) and H(0) in eq. 10.16 and 10.17 appear in our Big Bang model. Other values in other physically reasonable models will lead to similar results.

Since the changeover from a radiation-dominated phase to a matter-dominated phase is approximately estimated to be at:

Radiation – Matter Domination Transition: $t \cong 1.48 \times 10^{12}$ sec.

Due to uncertainties it may coincide with the Big Dip.

It seems reasonable to conclude the transition from radiation-dominated to matter-dominated causes the Big Dip to occur. The matter-dominated phase transition causes shrinkage as shown in a(t) in Fig. 10.4. *The universe contracts by one-third!* [60] We attribute the time delay between the transition and the Big Dip in a(t) to the time required for the transition to occur. (The universe is large at that time after all)

10.7.2 Universe Contraction – Early Massive Galaxies

The contraction would appear to "squeeze" the mass-energy in the universe giving it a "belly" (the Big Belly??? of "squeezed" mass-energy).This mass-energy contraction leads to the early formation of galaxies that disperse due to gravitation in the 13.5 Gyrs that follow. The subsequent expansion would also appear to create a "wake" similar to the wake of a water wave.

Evidence[61] has been found for the existence of a huge population of very massive galaxies (39+ have been found so far) that were created within one billion years after the Big Bang. This population of early galaxies is inconsistent with the standard present-day models of galaxy formation. The Big Dip occurs at 2.76 million years – well before one billion years – consistent with the formation of early massive galaxies.

A concentration of mass-energy due to the contraction of the universe appears to present a possible solution. Universe contraction was not considered in the creation of models of galaxy formation.

Another possible source of universe concentration (and voids) of energy appears in our Quantum Big Bang Model. The cause is a large difference in expansion rates (Hubble Constant variations) at the center of the Big Bang compared to the outer edge of the Big Bang.

10.7.3 Overshoot in H(t)

The result of the Radiation-Matter transition seems to be a negative H(t) for the energy density Ω_T. H(t) "overshoots" and becomes negative. Crudely put, the clumping of matter in the matter-dominated phase appears to introduce a compactness that results in a decrease in universe size and concentrations of energy.

10.7.4 Voids and Bubbles in Space after the Big Dip

The Big Dip concentrates mass-energy at the contraction. The subsequent expansion creates a "type" of wave that generates massive galaxies (bubbles of mass-energy), and also voids – bubbles of space devoid of galaxies. In the course of the

---

[60] Rather like the condensation of water vapor to liquid.
[61] T. Wang *et al*, Nature **572**, 211 (2019).

following thirteen or so billion years gravitation causes a dispersion of galaxies, voids and bubbles leading to the present day observed distribution.

10.7.5 Mystery of the Big Dip in H(t) - A Scenario

At the Big Dip H(t) changes from a declining to a rising trajectory. Based on this fact and the Big Bang model presented in Blaha (2019c) the following scenario seems reasonable:

1. The initial peak, and immediate decline, in H(t) is due to the Y black body radiation phase pressure that decreases rapidly after the Big Bang metastate ends. Thus Ω_T declines rapidly with the Y pressure decline. (Note Ω_T is a sum of energy density and pressure.)

2.   The peak in Ω_T reflects an influx of energy (from the Megaverse?) that causes H(t) to begin increasing. There is also a dip below zero in H(t) signifying the shrinkage of the universe as a(t).

3. Afterwards Ω_T continues to be significant and increasing as a(t) and H(t) rise to the present time.

4. In the future Ω_T should continue to rise. The energy increase that this situation implies suggests a certain reality to our Big_Bang-Quantum_Vacuum Theory. .

10.8 Radius of Big Bang Universe

The current radius of the universe is estimated[62] to be 4.314×10^{28} cm. Using the estimate in eq. 10.16 we find the radius of the Big Bang region at t = 0 is the not unreasonable value:

$$r_{\text{Big Bang}} = 1.376 \times 10^{-64} \text{ cm} \tag{10.29}$$

or

$$r_{\text{Big Bang}} = 8.532 \times 10^{-32}/M_{\text{Planck}}$$

10.9 Particles vs. Universes/Megaverses

How does an elementary particle differ from a universe or Megaverse particle? It seems that elementary particles are physically produced on-shell from on-shell particles. Universe and Megaverse Particles are produced by far off-shell, highly energetic initial particles.

---

[62] M. Tanabashi *et al*, Phys. Rev. **D98**, 030001 (2018).

10.10 A QED-like Vacuum Polarization Model of H(t)

The origin of the original model in a QED-like vacuum polarization (Appendix 10-A) and its extension based on similar considerations suggests that universes may well be types of particles as we picture in Octonion Cosmology. The same view may apply to Megaverses. They may also be types of particles.

If the vacuum polarization viewpoint is correct then it would suggest all universes have a similar pattern of growth within the framework of space 6 of Octonion Cosmology. It would also suugest that all Megaverses have a similar pattern of growth within the framework of space 5 of Octonion Cosmology.

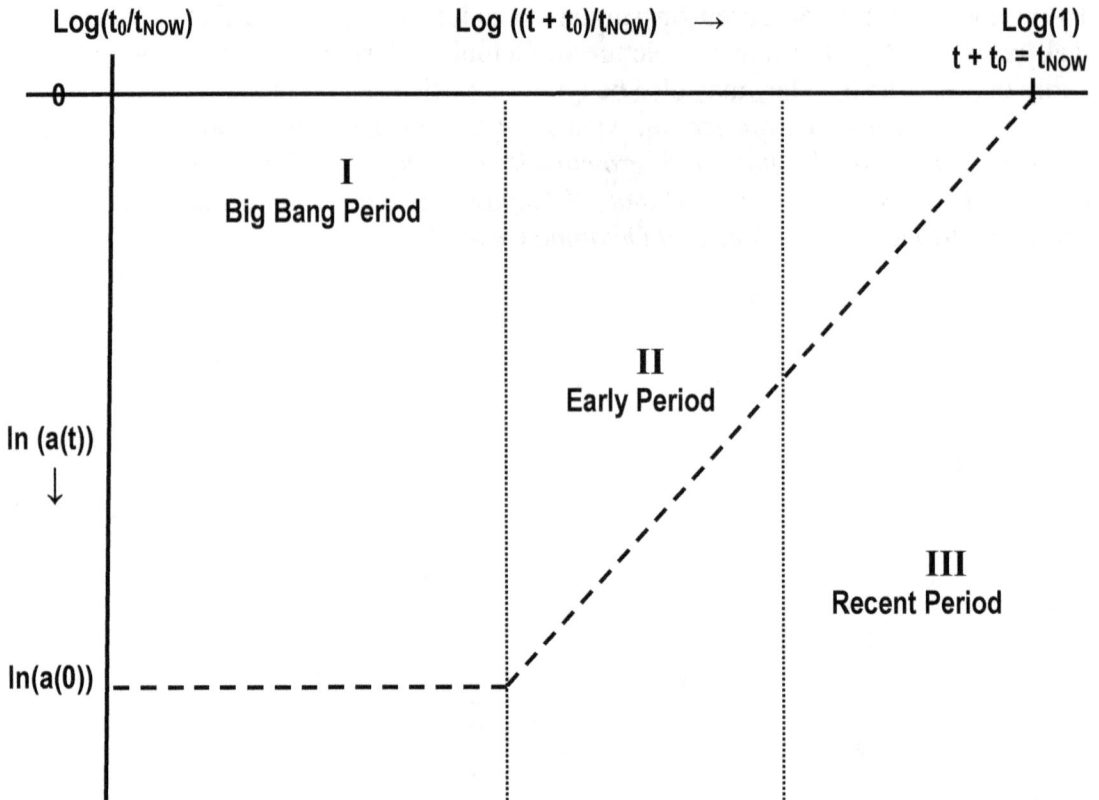

Figure 10.1. General form of a(t) of eq. 10.9 from t = 0 to t_{NOW} (the present,) It is not a straight line. The three periods are described below. Ln(a(0)) is a constant (and negative) since a(0) is non-zero.

10.11 Regions of a(t)

Our vacuum polarization-like model displayed in Fig. 10.1 has three regions in time:

I The "Near" Big Bang Period

a(t) in this period is based on the full expression in eq. 10.10 with h = 0.

II The Larger time Big Bang Period

a(t) in this period is based on the full expression in eq. 10.7 .

III The "Large" time Period

a(t) in this period is based on the full expression in eq. 22.1 above.

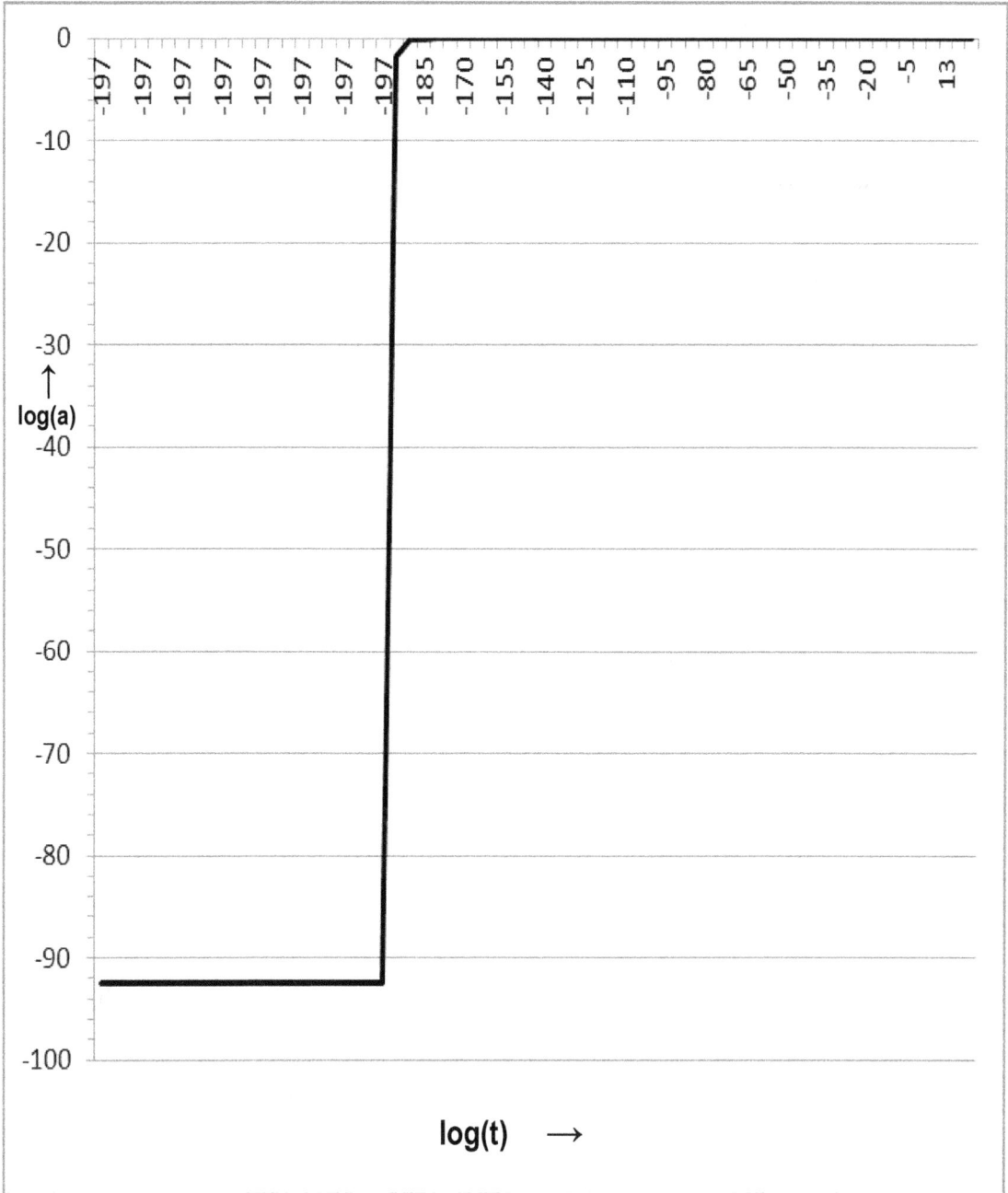

Figure 10.2. Plot of $\log_{10}(a(t))$ of eq. 10.9 from t = 0 to the present, t_{NOW}. Time (in s) and a(t) are plotted logarithmically to base 10: $\log_{10}(t)$ and $\log_{10}(a)$.. Note a(t) = 3.32495×10^{-93} until t = 10^{-200} s. After a short interval a(t) ranges from 0.87 to 1.

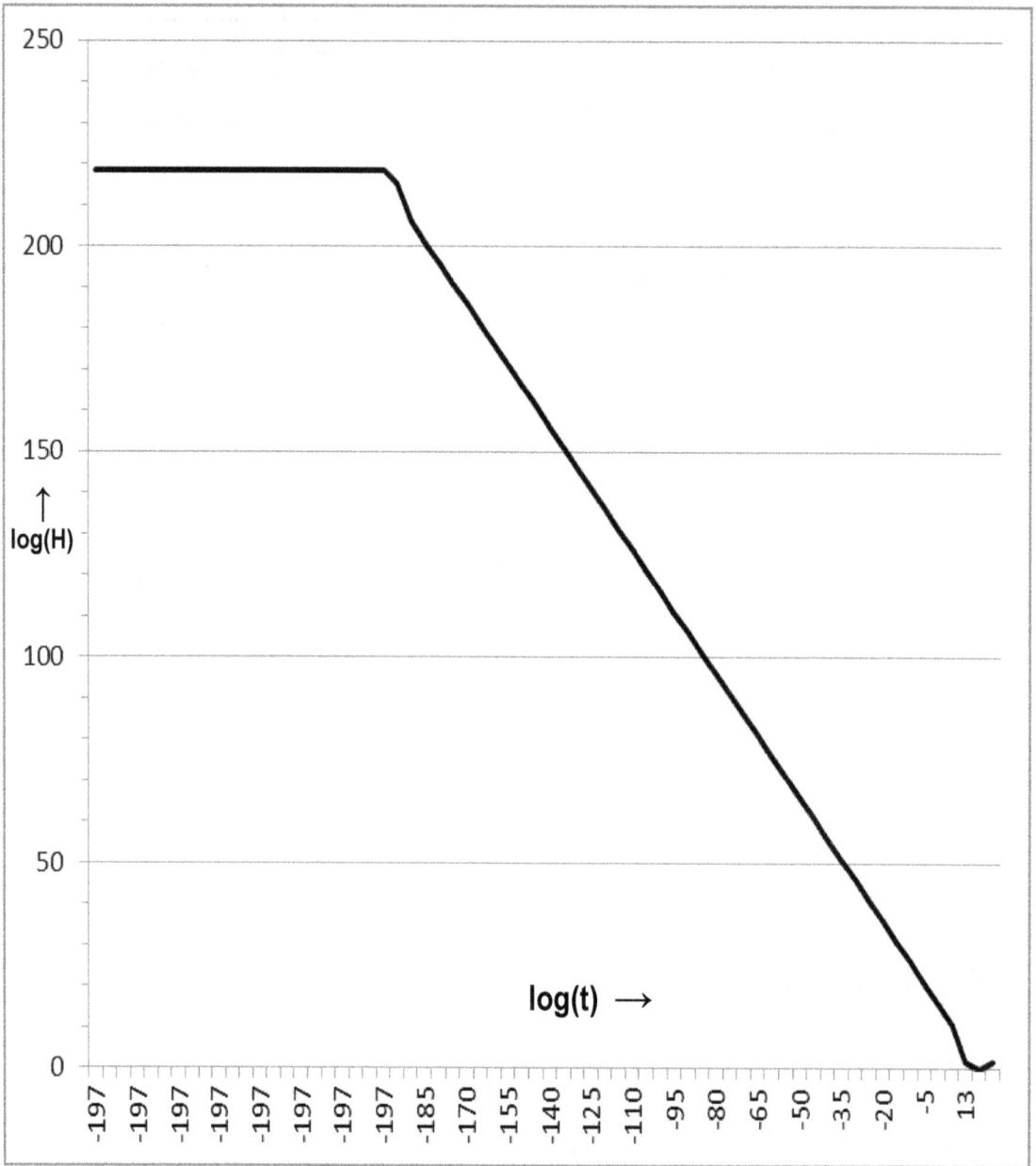

Figure 10.3. Plot of the log of the Hubble Parameter $\log_{10}(H(t))$ of eq. 10.10 from t = 0 to the present, t_{NOW}. Time (in s) and the Hublle Parameter (in km s^{-1} Mpc^{-1}) are plotted logarithmically to base 10: $\log_{10}(t)$ and $\log_{10}(H)$. Note H(t) = 3.5686×10^{218} km s^{-1} Mpc^{-1} until t = 10^{-200} s. Then H(t) ranges from 1.3×10^{215} km s^{-1} Mpc^{-1} to 68 km s^{-1}Mpc^{-1}.

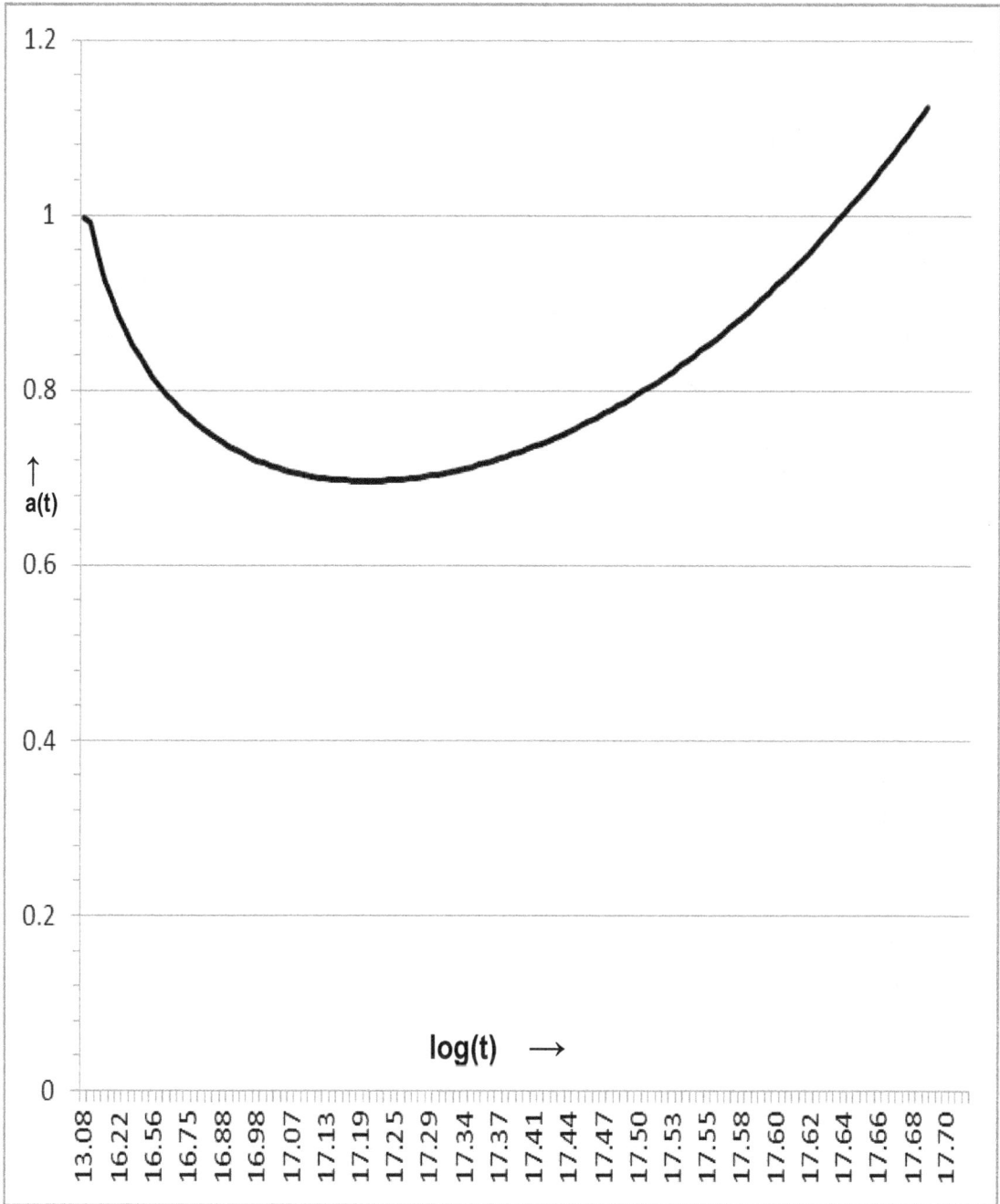

Figure 10.4. Plot of a(t) from eq. 10.9 vs. $\log_{10}(t)$ from t = 1.198×10^{13} s to t = 5.08×10^{17} s. Note a(t_{NOW}) = 1. The region of the minimum of a(t) corresponds to the Big Dip region of H(t).

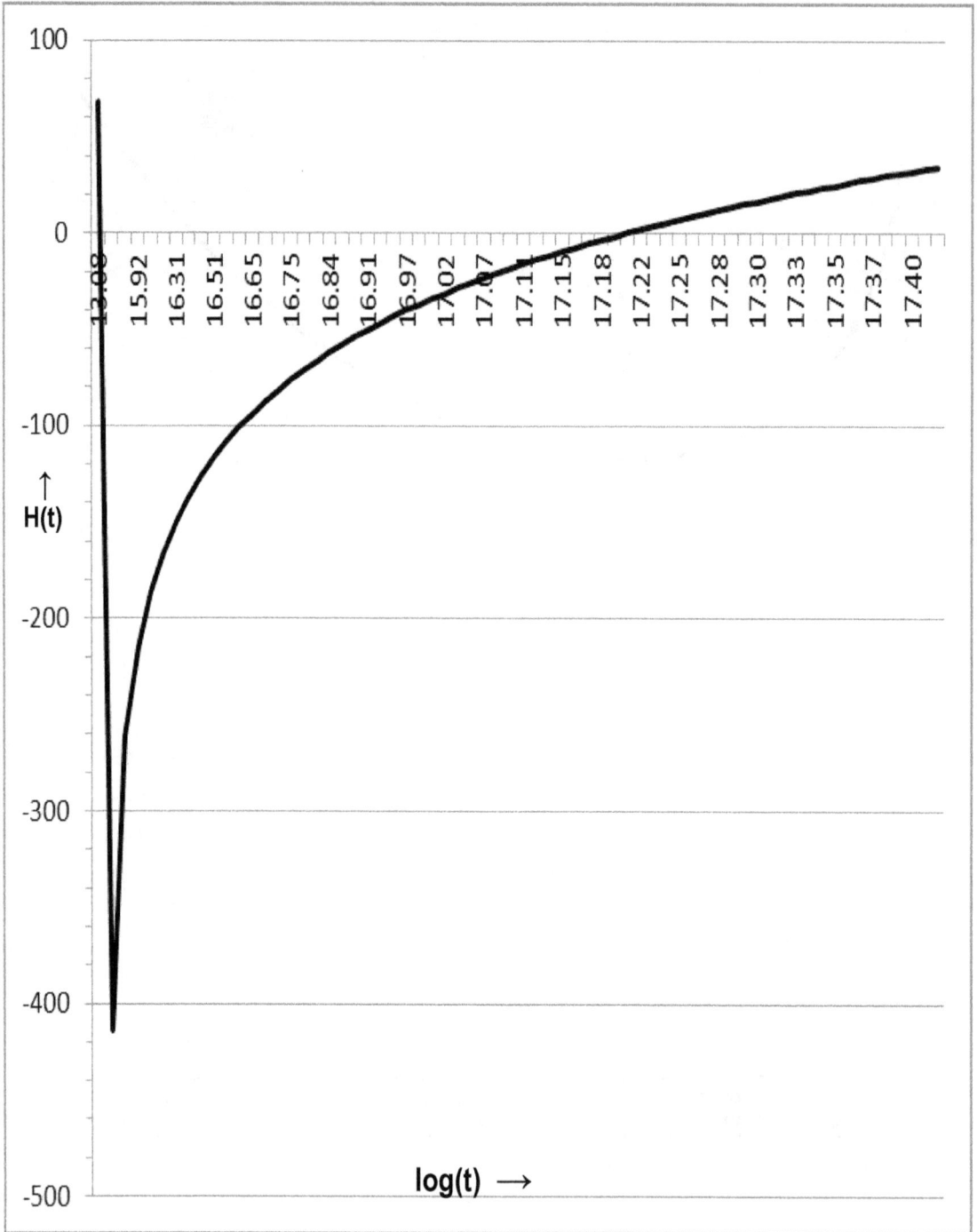

Figure 10.5. Plot of H(t) of eq. 10.10 vs. $\log_{10}(t)$ from t = 1.198×10^{13} to t = 5.08 $\times 10^{17}$ s. The Big Dip, the minimum of H(t), occurs at t = 4.1199×10^{14} s "shortly" after the radiation–matter transition in the universe.

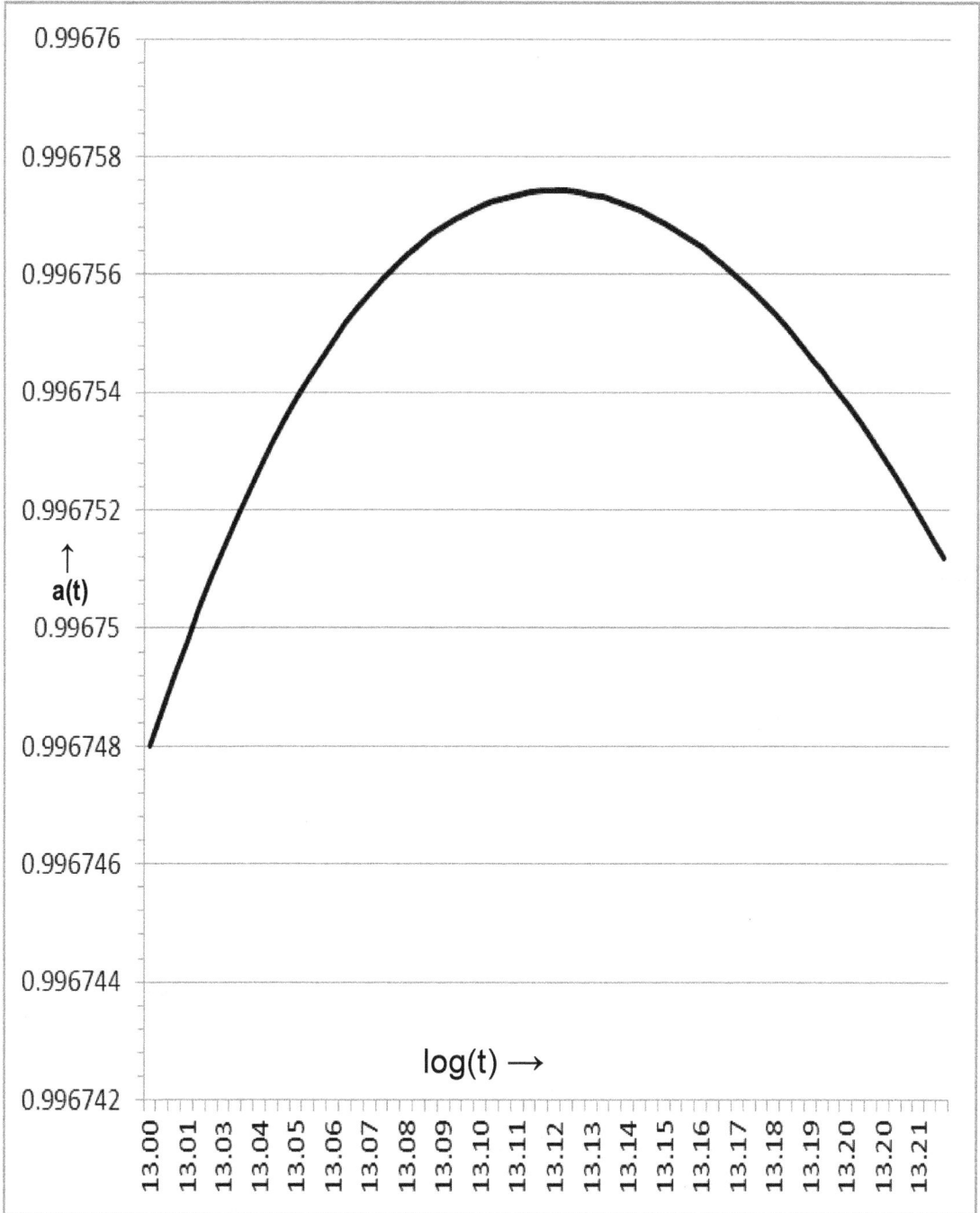

Figure 10.6. Plot of a(t) of eq. 10.9 vs. $\log_{10}(t)$ around the 380,000 year point from t = 1.01 × 10^{13} to t = 1.65 × 10^{13} s. The 380,000 year point corresponds to $\log_{10}(t)$ = 13.0786.

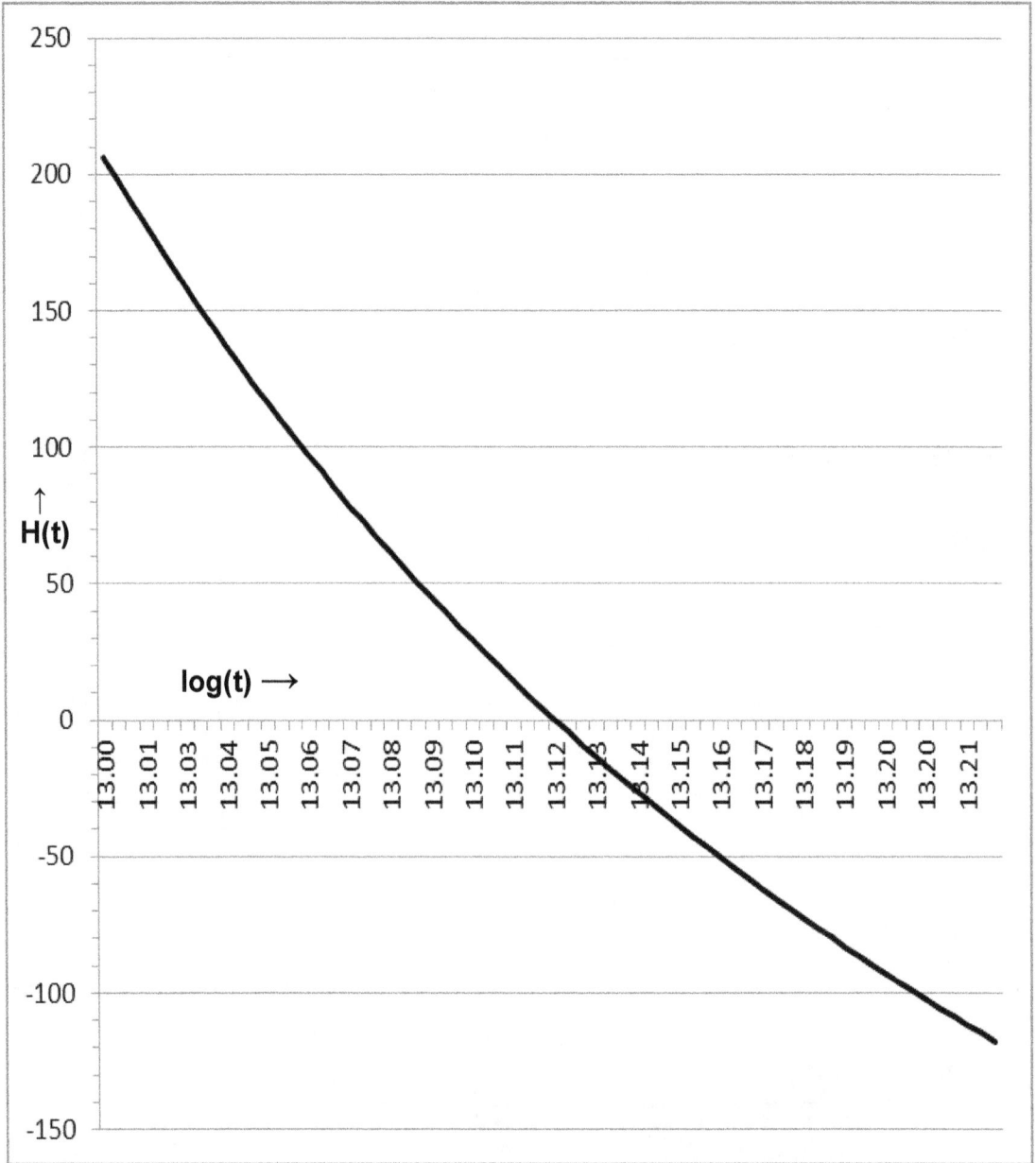

Figure 10.7. Plot of H(t) of eq. 10.10 vs. $\log_{10}(t)$ around the 380,000 year point from t = 1.01×10^{13} to t = 1.65×10^{13} s. The H(380000 years) value is 67.8 km s^{-1} Mpc^{-1}.

Appendix 10-A. Growth of Universes due to Vacuum Polarization

In this appendix[63] we show that the universal scale factor $a(t)$ of eq. 22.1[64]

$$a(t) = (t/t_{now})^{g + ht} \tag{22.1}$$

originates in the vacuum polarization of a spin 0 (boson) universe particle interacting with a new vector QED-like quantum field. Since the vector interaction would presumably affect all universe particles in the Megaverse, it appears that the universal scale factor would apply to all universes. The result would be a common universal pattern of evolution for all universes in the Megaverse.

10-A.1 Original Motivation for Relating the Scale Factor to Vacuum Polarization

It has become evident that the universal scale factor specified by eq. 22.1 leads to a physically reasonable scenario for the life history of a universe. The question that immediately arises is the cause of this form for $a(t)$. How does it work so well for both extremely early times near the Big Bang and for recent times as well?

We begin by expressing the scale factor as

$$a(t, T) = (t/T')^{g + ht} \tag{22.10}$$

where T is a time scale. The value of the exponents g and h are

$$h = 2.25983 \times 10^{-18} \tag{22.6}$$
$$g = 0.000282377 = 2.82377 \times 10^{-4}$$

For very early times "near" the Big Bang metastate time of $<10^{-165}$ sec we can approximate the universal scale factor with

$$a(t) \cong (t/T)^g \tag{10-A.1}$$

At first glance the value of g is an arbitrary constant set through the values of the Hubble Constant at $t = 380,000$ years and $T = t_{now}$. However the value of g is remarkably similar to the value of a renormalization exponent in the author's

---

[63] Some of the material in this appendix appears in Blaha (2019c) and (2019d).
[64] Equations label as "22." Appear in Blaha (2019e).

calculation of the Fine Structure Constant α in 1973 and in section 5.5.3 of Blaha (2019b).

The value of the Fine Structure Constant was based entirely on vacuum polarization in massless Quantum Electrodynamics (QED) – a vacuum effect of electromagnetism. Although astrophysicists do not think of the Big Bang and the expansion of the universe as a vacuum effect, it is clear from the plots shown earlier that universe growth is dependent on its energy density and an influx (appearance) of energy from somewhere (the Megaverse in our view). Thus the growth of the universe is directly dependent on the vacuum (of the Megaverse).

In massless QED we found that the vacuum polarization had the form:[65]

$$F_1(\alpha)(p/\Lambda)^{2g_{QED}} \tag{10-A.2}$$

where $F_1(\alpha)$ is the "eigenvalue function" for the Fine Structure Constant[66] of the Johnson-Baker-Willey model of massless QED, p is the momentum, and Λ is the ultraviolet cutoff. The value of g_{QED} that corresponded to the Fine Structure Constant is

$$g_{QED} = -0.0005805375 \tag{10-A.3}$$

and the Fine Structure Constant was correctly found (well within experimental limits) to be

$$\alpha_{calculated}(g_{QED}) = 0.007297353 \tag{10-A.4}$$

to 9 digit accuracy according to the Particle Data Table of 2018.

Comparing our g value (eq. 22.6 above) with g_{QED} we

$$-g \cong -\tfrac{1}{2}\, g_{QED} \tag{10-A.5}$$

Excepting the factor of two, we find a remarkable numerical coincidence comparing electron vacuum polarization with the universe scale factor where high momentum electron polarization corresponds to very early universe time. This coincidence is not accidental.

10-A.2 A New Vector Interaction for Universe Particles

We assume universes can be treated as particles in 4-dimensional space-time.[67] Since experiments appear to have shown that the universe does not rotate (does not have spin)[68] we will assume the universe is a spin 0 boson. We assume that universes have a

---

[65] Eq. 12 in S. Blaha, Phys Rev **D9**, 2246 (1973).

[66] The author calculated α exactly (within current experimental limits) in Blaha (2019a) and (2019b).

[67] Universes are composite entities but we can treat them as quantum particles in the same manner as physicists treated protons and neutrons etc. as quantum particles before quark theory was accepted.

[68] The lack of universe rotation (spin) is indicated by a study of Cosmic Microwave Background (CMB) by D. Saadeh *et al*, Phys. Rev. Lett. **117**, 313302 (2016).

vector field interaction similar to QED. It is possible that the quantum vector $Y^\mu(x)$ field of the Big Bang quantum coordinates, treated earlier, may be the vector field universe interaction field as well.

Given this QED-like framework, universe-antiuniverse pair production and vacuum polarization becomes possible. We assume the QRD-like lagrangian

$$\mathcal{L} = \tfrac{1}{2}\,(\partial_\mu\varphi^\dagger\partial^\mu\varphi - m^2\varphi^\dagger\varphi) - ie_0 \colon \varphi^\dagger(\overrightarrow{\partial_\mu} - \overleftarrow{\partial_\mu})\,\varphi \colon A^\mu + e_0^2 \colon A^2 \colon \colon \varphi^\dagger\varphi \colon + \delta m^2 \colon \varphi^\dagger\varphi \colon$$

(10-A.6)

where $\varphi(x)$ is a "charged" quantum universe particle field.[69]

We now proceed to calculate the second order vacuum polarization of a universe particle.

10-A.3 Second Order Vacuum Polarization of a Universe Particle

The one loop vacuum polarization Feynman diagram is

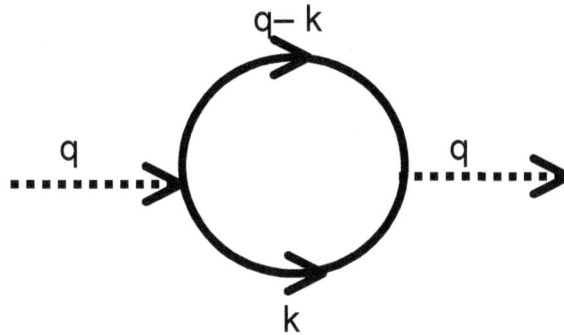

Figure 10-A.1 One loop vacuum polarization boson Feynman diagram.

We evaluate

$$I_{\mu\nu} = (-ie_0)^2 \int \frac{d^4k}{(2\pi)^4}\, \frac{i}{(k^2 - m^2 + i\varepsilon)}\, \frac{i}{(k^2 - m^2 + i\varepsilon)}\,(q - 2k)_\mu (q - 2k)_\nu \qquad (10\text{-A.7})$$

$$= \frac{\alpha}{2\pi} \int_0^\infty dz_1 \int_0^\infty dz_2\, \frac{g_{\mu\nu}\,\exp[i(q^2 z_1 z_2/(z_1 + z_2) - (m^2 + i\varepsilon)\,(z_1 + z_2))]}{(z_1 + z_2)^3} + \text{gauge terms}$$

upon introducing parameters z_1 and z_2 to enable exponentiation and integration over k, where

$$\alpha = e_0^2/4\pi \qquad (10\text{-A.8})$$

---

[69] The charge is not electromagnetic charge.

Applying $q^2 \partial/\partial q^2$ to $I_{\mu\nu}$ to eliminate the quadratic divergent part, and then using the identity

$$1 = \int\limits_0^\infty d\lambda/\lambda \; \delta(1 - (z_1 + z_2)/\lambda)$$

and letting $z_i = \lambda x_i$ we obtain

$$I_{\mu\nu} = \frac{i\,\alpha}{2\pi} \; q^2 g_{\mu\nu} \int dx_1 \int dx_2 \int d\lambda/\lambda \; x_1 x_2 \exp[i\lambda(q^2 x_1 x_2 - (m^2 + i\varepsilon))] \; \delta(1 - x_1 - x_2)$$

(10-A.9)

up to gauge terms. The λ integration yields a logarithmic divergence which we cut off.
    Then

$$I_{\mu\nu} = \frac{i\,\alpha}{2\pi} \; q^2 g_{\mu\nu} \int\limits_0^1 dx \; x(1-x) \ln(q^2 x(1-x) - m^2) \qquad (10\text{-A.10})$$

which becomes

$$I_{\mu\nu} = \frac{i\,\alpha}{12\pi} \; q^2 g_{\mu\nu} \ln(\Lambda^2/m^2) + \dots \qquad (10\text{-A.11})$$

with finite terms not shown.
Thus we find the renormalization constant Z_3 for the scalar universe particle case to be

$$Z_3 = 1 - \alpha/12\pi \; \ln(\Lambda^2/m^2) \qquad (10\text{-A.12})$$

If we let

$$\alpha_U = \alpha/4 \qquad (10\text{-A.13})$$

then we obtain the form similar to the one loop value of Z_3 for spin ½ electron QED:

$$Z_3 = 1 - \alpha_U/3\pi \; \ln(\Lambda^2/m^2) \qquad (10\text{-A.14})$$

We now provisionally assume that α is the QED fine structure constant. We denote it as α_{QED}. Note $\alpha_{QED} = 0.007297353$. (Later we will show that the value of $g = 0.000282377$ that we found earlier follows from this choice.)
    Thus the "fine structure constant" α_U for our vector interaction is

$$\alpha_U \equiv \alpha_{QED}/4 = 0.001824338 \qquad (10\text{-A.15})$$

We now turn to the Johnson-Baker-Willey (JBW) model of massless QED since at ultra high energy our vector interaction theory with lagrangian eq. 10-A.6 becomes the JBW model. In the JBW model we calculated α_{QED} and found the corresponding power which we denote g_{QED}. Now we perform the same calculation and find the g value denoted g_U corresponding to α_U. The value of g_U will be seen to lead to the power g in the universal scale factor almost exactly.

10-A.4 Massless QED-like Calculation of Vacuum Polarization to All Orders

We calculated[70] an approximate solution to all orders of the four divergences in the JBW model:

Z_1	- vertex renormalization factor
$Z_2 = Z_1$	- self-energy renormalization factor
Z_3	- vacuum polarization renormalization factor
δm	- self-mass renormalization

The renormalization constants appear in the expressions:

$$S_F(p) = Z_2 S'_F(p)$$
$$D_F(q)_{\mu\nu} = Z_3 D'_F(q)_{\mu\nu}$$
$$V_\mu(q', q) = Z_1^{-1} V'_\mu(q', q)$$

where $S'_F(p)$, $D'_F(q)_{\mu\nu}$, and $V'_\mu(q', q)$ are the divergence-free QED physical propagators and vertex.

The JBW model was an attempt to eliminate the divergences of QED at very high energies where the electron mass may be neglected. It was extended to allow for non-zero electron mass.

The massless QED eigenvalue function of the JBW model was found in a series of papers and summarized in some detail by the paper of K. Johnson and M. Baker, Phys. Rev. **D8**, 1110 (1973) In this section we briefly outline the steps leading to the JBW eigenvalue function:

1. The electron self-energy and the photon vacuum polarization are calculated using the free electron and photon propagators.

2. The apparent quadratic divergence is reduced to a logarithmic divergence in the vacuum polarization.

3. The logarithmic divergence in the vacuum polarization is further reduced to a single power of the logarithm of the ultraviolet cutoff denoted Λ.

4. The coefficient of the divergent logarithmic term is denoted

$$(x/2\pi)F(x)$$

where x is the bare fine structure constant α_0.

---

[70] See S. Blaha, Phys. Rev. **D9**, 2246 (1974). Reprinted in Appendix F. Blaha (2019b) presents the JBW model and the calculation of α_{QED}.

5. If

$$F(x_0) = 0$$

for some value x_0 then a consistent divergence-free (finite) solution of massless QED is found.

6. The $F(x)$ function is reduced to the eigenvalue function $F_1(x)$ which is the sum of logarithmic divergences of all single closed electron loop diagrams. If the eigenvalue function $F_1(x)$ has a zero at x_0 then $F(x_0) = 0$. Consequently the eigenvalue condition becomes

$$F_1(x_0) = 0$$

7. Adler[71] made the important observation that a zero F_1 would necessarily be an essential singularity:

$$d^n F_1(x)/dx^n|_{x=x_0} = 0 \qquad \text{for all } n > 0$$

Depending on the summation in perturbation theory of the relevant vacuum polarization diagrams might may occur at the bare coupling constant α_0 or the physical coupling constant $\alpha = 1/137$ ...

8. This author (See Appendix F of Blaha (2019e).) calculated an approximation to F_1 which did not explicitly display an essential singularity and did not have a zero at the physical fine structure constant α.

10-A.5 Blaha's Approximate Calculation of the Eigenvalue Function

In 1974 this author[72] formulated an approximation to the equations of massless QED and solved them for the vacuum polarization, electron self-energy and the vertex renormalization. The approximation is described in detail in the author's Phys. Rev. D paper in Appendix F.

The approximate solution for $F_1(x)$ had the encouraging feature that it reproduced the known[73] low order exact calculations of $F_1(x)$:

$$F_{1 \text{ low order}}(x) = 2/3 + x/(2\pi) - (1/4)[x/(2\pi)]^2$$

The *approximate* solution, which summed pieces of the vacuum polarization given by the diagrams of Figs. 2 and 3 in Appendix F, yields the algebraic equations:[74]

---

[71] S. Adler, Phys. Rev. **D5**, 3021 (1972).

[72] S. Blaha, Phys Rev **D9**, 2246 (1973).

[73] J. Rosner, Phys. Rev. Lett. **17**, 1190 (1966).

[74] Blaha *op. cit.* The solution for the eigenvalue function is clearly best expressed in terms of the g factor in the exponents of the divergent renormalization factors.

$A_1 = (g + 1)(1 - 2g^2)/[(g + 2)(g - 1)]$

$A_2 = [8g^2(2g + 1) - (2g^3 + 2g^2 + g - 2)(g^2 + 2g + 2)]/[2(g^2 - 1)(g^2 - 4)]$

$A_3 = -2(1 + 3g + 6g^2 + 2g^3)/[g(g + 1)]$

$A_4 = -(g + 2)(1 + 5g + 6g^2 + 2g^3)/[g(g^2 - 1)] - 1/(g + 1)$

$\psi = [gA_3 - (4 + 2g)A_1]/[(4 + 2g)A_2 - g A_4]$

$(\alpha/2\pi) = [gA_4 - (4 + 2g)A_2]/(A_4A_1 - A_2A_3)$ (10-A.16)

$F_1(g) = (2/3)(1 - 3g^2/2 - g^3) - (\alpha/4\pi)[(2 + 4g + 4g^2)(g - 2) + \alpha\psi g^3]/[(g^2 - 1)(g - 2) +$
$+ \alpha(2 + 4g + 4g^2)(g - 2) + \alpha\psi g^3]$

as a function[75] of $g = g_{QED}$ with ψ specifying the gauge, and with the definitions

$$\Gamma_\mu(p) = f(\gamma_\mu + 2g\gamma \cdot pp_\mu/p^2)(p/\Lambda)^{2g}$$ (10-A.16')
$$S_F = [f\gamma \cdot p(p/\Lambda)^{2g}]^{-1}$$
$$\Gamma_{\mu\alpha}(p) = (f_3/p^2)(\gamma \cdot p\gamma_\mu\gamma_\alpha - \gamma_\alpha\gamma_\mu\gamma \cdot p)(p/\Lambda)^{2g}$$

and

$$F_1 = (2/3)(1 - 3g^2/2 - g^3) - f_3/f$$

in the notation of Appendix F.[76] Eqs. 5.4 and 5.5 manifestly cannot lead to a form of F_1 with an essential singularity due to their algebraic form.

The plot of F_1 did not show a zero of F_1 at the physical fine structure constant. Thus the hopes raised by the JBW model seemed dashed—at least in our approximate solution *then*. Later in Blaha (2019a) and (2019b) we revived the hope of a satisfactory eigenvalue function with an eigenvalue at the physical value of the fine structure constant α.

As pointed out in our 1974 Phys. Rev. D paper the eigenvalue function F_1 does not have a zero at the known value of the Fine Structure Constant.

In 2019 we reconsidered our approximation and found a value of α very near the measured value. We defined

$$F_2(\alpha) = F_1(\alpha) - [2/3 + \alpha/(2\pi) - (1/4)[\alpha/(2\pi)]^2]$$ (10-A.16'')

10-A.6 The *Correct* Value of the QED Fine Structure Constant

We found a neighborhood in the range of g where $F_2(\alpha) \approx 0$ with an approximate value for the known fine structure constant. We reconsidered our calculation recently and found

---

[75] We use $F_1(g)$ and $F_1(\alpha(g))$ interchangeably.
[76] Blaha (2019e).

$$g_{QED} = -0.000580537 \qquad (10\text{-A}.17)$$

and

$$\alpha_{QED} = 0.007297353 \qquad (10\text{-A}.18)$$

which gives the QED fine structure constant (an irrational number) accurately to 9 places.

10-A.7 Extension of Calculation to Non-Abelian Interactions

In Blaha (2019b) we generalized eq. 10-A.16 to the cases of the Weak interaction and the Strong interaction coupling constants by inserting a group theoretic factor in eq. 10-A.16:

$$c_G^{-1} = [(11/3)C_{ad} - 2C_f/3]/(16\pi)^3 \qquad (10\text{-A}.19)$$
$$(\alpha_G/2\pi) = c_G^{-1}[gA_4 - (4 + 2g)A_2]/(A_4A_1 - A_2A_3) \qquad (10\text{-A}.20)$$

where C_{ad} is the dimension of the fundamental representation of the non-abelian group and C_f is the number of fermions (fermion flavor) of the interaction.

We found good approximations to the SU(2) and SU(3) coupling constants.

10-A.8 Application of the Approximate Coupling Constant Calculation to Universe Vacuum Polarization

We now extend eq. 10-A.16 to the case of the vacuum polarization of a universe particle described by

$$(\alpha_U/2\pi) = [g_U A_4(g_U) - (4 + 2g_U)A_2(g_U)]/(A_4(g_U)A_1(g_U) - A_2(g_U)A_3(g_U)) \qquad (10\text{-A}.21)$$

since the second order form of Z_3 (eq. 10-A.14), which is the same as the QED second order form of Z_3, generalizes to all orders as a function of α_U.

Given the value of

$$\alpha_U = 0.001824338 \qquad (10\text{-A}.22a)$$

in eq. 10-A.15 we can extract the value of g_U by inverting eq. 10-A.21 to obtain:

$$g_U = -0.00014525 \qquad (10\text{-A}.22b)$$

10-A.9 Relating Vacuum Polarization to the Universal Scale Factor

We now relate the vacuum polarization found above to the growth of the universe as given by the universal scale factor. Eq. 10-A.16'

$$\Gamma_{\mu\alpha}(p) = (f_3/p^2)(\gamma \cdot p\gamma_\mu\gamma_\alpha - \gamma_\alpha\gamma_\mu\gamma \cdot p)(p/\Lambda)^{2g_U}$$

gives the vacuum polarization factor

$$\Gamma(p) = (p/\Lambda)^{2g_U} \qquad (10\text{-A}.23)$$

where

$$g_U = g/4 = -0.00014543 \qquad (10\text{-A}.24)$$

We now fourier transform $\Gamma(p)$ to coordinate space – in particular to time t[77]

$$a(t) = (1/2\pi) \int_0^{\infty} dp \, \exp(-ipt) \, \Gamma(p) \qquad (10\text{-A}.25)$$

$$= k \, (t/T)^{-2g_U} \qquad (10\text{-A}.26)$$

where k is a constant and where

$$1/T = \Lambda \qquad (10\text{-A}.27)$$

with Λ being the "momentum space" cutoff mass. From eq. 10-A.22b and 10-A.26 we find

$$g = -2g_U$$
$$= 0.0002905 \qquad (10\text{-A}.28)$$

From eq. 22.6 for the power g of a(t) we see

$$g = 0.000282377$$

Thus the value of g calculated from the vacuum polarization differs from the actual value of g by less than 3%. Given the approximate nature of our JBW calculation of vacuum polarization the agreement is remarkable.[78]

The dependence of the universal scale factor on g governs the small time behavior of the universe. Correspondingly, the dependence of the vacuum polarization on g_U is a large momentum phenomena. The parameter h in a(t) is set primarily by the large t (recent times) behavior of a(t). It corresponds to the infrared behavior of its fourier transform $\Gamma(p)$ when the infrared (possibly mass dependent) behavior of the vacuum polarization is calculated.

The preceding discussion demonstrates that our earlier assumption

$$\alpha_U \equiv \alpha_{QED}/4 = 0.001824338 \qquad (10\text{-A}.15)$$

---

[77] Those who might object to forier transforming to time t should remember that inside a Black Hole the "time-like" coordinate is the radius and the time variable t is comparable to a spatial coordinate. The possibility that the universe is a Black Hole is not excluded.

[78] And may be exact! The value of the Hubble Constant H in recent times varies from about 70 – 75 making the calculation of g also approximate. We chose an average value of 73.24 to obtain the value of g above. If we chose the current value for H to be 75.58 we would have $g = -2g_U$ exactly (eq. 25.28). Note: studies of binary black hole merger gravity waves have given a Hubble Constant of 75.2 km s^{-1} Mpc^{-1} (and earlier of 78 km s^{-1} Mpc^{-1}), and studies of light bent by distant galaxies give H = 72.5 km s^{-1} Mpc^{-1}. Thus the value H = 75.58 is not unreasonable. See section 22.1 for a summary of studies of H.

is correct since it leads directly to the power g in the universal scale factor. Thus the evolution of our universe is set by the vacuum polarization originating in the new vector interaction that we have introduced.

Using our notation we can rewrite eq. 10-A.6 as

$$\mathscr{L} = \tfrac{1}{2}\,(\partial_\mu\varphi^\dagger\partial^\mu\varphi - m^2\varphi^\dagger\varphi) - ie_{0U}\!:\varphi^\dagger(\overrightarrow{\partial_\mu} - \overleftarrow{\partial_\mu})\,\varphi\!:\,Y^\mu + e_{0U}{}^2\!:Y^2\!:\,:\varphi^\dagger\varphi\!: + \delta m^2\!:\varphi^\dagger\varphi\!:$$

$$(10\text{-A}.29)$$

The Y^μ vector field that we used to stabilize the Big Bang with quantum coordinates may now have a role as the interaction between universe particles. Incidentally we have also shown the viability of viewing universes as quantum particles just as nucleons were viewed as quantum particles before the acceptance of a quark substructure.

10-A.10 Proof that Our Universe's Growth Pattern is due to a Form of Vacuum Polarization

The calculation of g from the vacuum polarization of a vector field proves that the growth pattern of our universe is governed by vacuum polarization although it takes the form of depending on Dark Energy. It appears that Dark Energy *per se* does not exist except as a representation of the vacuum polarization generated by our vector gauge field theory. Note no direct physical evidence of Dark Energy interacting with "normal" matter exists. Dark Energy is inferred and may be vacuum polarization energy due to a new universe interaction.

Since the universal scale factor implies enormous, varying Dark Energy Ω_T as shown elsewhere, we view the growth of our universe as a vacuum polarization phenomenon.

10-A.11 Vacuum Polarization Interpretation of the Universal Scale Factor

The vacuum polarization view of the time evolution of the universe requires that we view the entire time evolution of the universe as a whole. Normally one views time as increasing. Feynman suggested we could also view time as flowing backward.

We now have a new view where we freeze the life history of the universe as a static event rather like a time lapse picture of a flower's growth. The thought process is similar to that of Feynman path integral formulations, which consider the complete path in time of a process.

10-A.12 Path Integral Formulation of a *Quantum* Scale Factor

It is possible define a path integral formalism for a *quantum* a(t) starting from the Friedmann equation and its lagrangian formulation. The Friedmann equation

$$d^2a/dt^2 + 4\pi G/3\,(\rho + 3p/c^2)\,a - \Lambda c^2 a/3 = 0$$

clearly resembles the Schrödinger equation. The sum over paths of a(t) would then have a path corresponding to the "classical" solution eq. 22.1 that we considered. Thus the complete "history" in a(t) (and its vacuum polarization equivalence) would be understandable. A potential benefit of the quantum a(t) is the elimination of divergences at t = 0 by quantum smearing.

10-A.13 Scale Factors of Other Universes

The considerations of a vector interaction generating the universal scale factor of our universe also applies to other universes in the Megaverse due to the assumption of a common Y interaction for all universes. Thus the growth pattern of our universe applies to other universes. It is a general feature of universes.

Other universes have the same evolutionary development as our universe.

10-A.14 Universe Eigenvalue Function

The eigenvalue function for universes is based on the general eigenvalue function

$$F_2(\alpha) = F_1(\alpha) - [2/3 + \alpha/(2\pi) - (1/4)[\alpha/(2\pi)]^2] \qquad (10\text{-A.16''})$$

The universe eigenvalue function is

$$F_2(\alpha_U) = F_1(\alpha_U) - [2/3 + \alpha_U/(2\pi) - (1/4)[\alpha_U/(2\pi)]^2] \qquad (10\text{-A.30})$$

If

$$F_2(\alpha_U) = 0 \qquad (10\text{-A.31})$$

as it does for

$$\alpha_U \equiv \alpha_{QED}/4 = 0.001824338 \qquad (10\text{-A.15})$$

where we find

$$F_2(\alpha_U = 0.001824338) = 5.10824 \times 10^{-12}$$

then, given the approximation used, $F_2(\alpha_U = 0.001824338)$ is essentially zero.

Just as a zero of F_2 implies quasi-free particle behavior at high energy in the JBW model we find universe particles are quasi-free.[79] (Gravitation between universes still exists.) See Appendix D of Blaha (2019e) for the case of free spin ½ universes.

10-A.15 A Hierarchy of Interactions

The vector interaction which we denote with the quantum field label Y with $e_U = (4\pi\alpha_U)^{1/2}$, and the other coupling constants for QED, Weak SU(2) and Strong SU(3) have a remarkable regularity—they double from interaction to interaction as Fig. 10-A.2 shows.

The deeper significance of this regularity is not known.

---

[79] Universe particles do have a low order divergent piece (exhibited in eq. 25.16'') in their vacuum polarization as shown in Blaha (2019b).

INTERACTION	COUPLING CONSTANT[80]
Y Interaction e_U	0.152
QED $e_{QED} = (4\pi\alpha_{QED})^{\frac{1}{2}}$	0.303
Weak SU(2) g_W	0.619
Strong SU(3) g_S	1. 22

Figure 10-A.2. The interaction constants show a regular doubling. The cause of the doubling is not apparent.

[80] M. Tanabashi *et al* (Particle Data Group), Phys. Rev. D**98**, 030001 (2018).

REFERENCES

Akhiezer, N. I., Frink, A. H. (tr), 1962, *The Calculus of Variations* (Blaisdell Publishing, New York, 1962).

Bjorken, J. D., Drell, S. D., 1964, *Relativistic Quantum Mechanics* (McGraw-Hill, New York, 1965).

Bjorken, J. D., Drell, S. D., 1965, *Relativistic Quantum Fields* (McGraw-Hill, New York, 1965).

Blaha, S., 1995, *C++ for Professional Programming* (International Thomson Publishing, Boston, 1995).

_____, 1998, *Cosmos and Consciousness* (Pingree-Hill Publishing, Auburn, NH, 1998 and 2002).

_____, 2002, *A Finite Unified Quantum Field Theory of the Elementary Particle Standard Model and Quantum Gravity Based on New Quantum Dimensions™ & a New Paradigm in the Calculus of Variations* (Pingree-Hill Publishing, Auburn, NH, 2002).

_____, 2004, *Quantum Big Bang Cosmology: Complex Space-time General Relativity, Quantum Coordinates™ Dodecahedral Universe, Inflation, and New Spin 0, ½, 1 & 2 Tachyons & Imagyons* (Pingree-Hill Publishing, Auburn, NH, 2004).

_____, 2005a, *Quantum Theory of the Third Kind: A New Type of Divergence-free Quantum Field Theory Supporting a Unified Standard Model of Elementary Particles and Quantum Gravity based on a New Method in the Calculus of Variations* (Pingree-Hill Publishing, Auburn, NH, 2005).

_____, 2005b, *The Metatheory of Physics Theories, and the Theory of Everything as a Quantum Computer Language* (Pingree-Hill Publishing, Auburn, NH, 2005).

_____, 2005c, *The Equivalence of Elementary Particle Theories and Computer Languages: Quantum Computers, Turing Machines, Standard Model, Superstring Theory, and a Proof that Gödel's Theorem Implies Nature Must Be Quantum* (Pingree-Hill Publishing, Auburn, NH, 2005).

_____, 2006a, *The Foundation of the Forces of Nature* (Pingree-Hill Publishing, Auburn, NH, 2006).

_____, 2006b, *A Derivation of ElectroWeak Theory based on an Extension of Special Relativity; Black Hole Tachyons; & Tachyons of Any Spin.* (Pingree-Hill Publishing, Auburn, NH, 2006).

_____, 2007a, *Physics Beyond the Light Barrier: The Source of Parity Violation, Tachyons, and A Derivation of Standard Model Features* (Pingree-Hill Publishing, Auburn, NH, 2007).

_____, 2007b, *The Origin of the Standard Model: The Genesis of Four Quark and Lepton Species, Parity Violation, the ElectroWeak Sector, Color SU(3), Three Visible Generations of Fermions, and One Generation of Dark Matter with Dark Energy* (Pingree-Hill Publishing, Auburn, NH, 2007).

_____, 2008a, *A Direct Derivation of the Form of the Standard Model From GL(16) (Pingree-Hill Publishing, Auburn, NH, 2008).*

_____, 2008b, *A Complete Derivation of the Form of the Standard Model With a New Method to Generate Particle Masses Second Edition* (Pingree-Hill Publishing, Auburn, NH, 2008)

_____, 2009, *The Algebra of Thought & Reality: The Mathematical Basis for Plato's Theory of Ideas, and Reality Extended to Include A Priori Observers and Space-Time Second Edition* (Pingree-Hill Publishing, Auburn, NH, 2009).

_____, 2010a, *Operator Metaphysics: A New Metaphysics Based on a New Operator Logic and a New Quantum Operator Logic that Lead to a Mathematical Basis for Plato's Theory of Ideas and Reality* (Pingree-Hill Publishing, Auburn, NH, 2010).

_____, 2010b, *The Standard Model's Form Derived from Operator Logic, Superluminal Transformations and GL(16)* (Pingree-Hill Publishing, Auburn, NH, 2010).

_____, 2010c, *SuperCivilizations: Civilizations as Superorganisms* (McMann-Fisher Publishing, Auburn, NH, 2010).

_____, 2011a, *21st Century Natural Philosophy Of Ultimate Physical Reality* (McMann-Fisher Publishing, Auburn, NH, 2011).

_____, 2011b, *All the Universe! Faster Than Light Tachyon Quark Starships & Particle Accelerators with the LHC as a Prototype Starship Drive Scientific Edition* (Pingree-Hill Publishing, Auburn, NH, 2011).

_____, 2011c, *From Asynchronous Logic to The Standard Model to Superflight to the Stars* (Blaha Research, Auburn, NH, 2011).

_____, 2012a, *From Asynchronous Logic to The Standard Model to Superflight to the Stars volume 2: Superluminal CP and CPT, U(4) Complex General Relativity and The Standard Model, Complex Vierbein General Relativity, Kinetic Theory, Thermodynamics* (Blaha Research, Auburn, NH, 2012).

_____, 2012b, *Standard Model Symmetries, And Four And Sixteen Dimension Complex Relativity; The Origin Of Higgs Mass Terms* (Blaha Reasearch, Auburn, NH, 2012).

_____, 2013a, *Multi-Stage Space Guns, Micro-Pulse Nuclear Rockets, and Faster-Than-Light Quark-Gluon Ion Drive Starships* (Blaha Research, Auburn, NH, 2013).

_____, 2013b, *The Bridge to Dark Matter; A New Sister Universe; Dark Energy; Inflatons; Quantum Big Bang; Superluminal Physics; An Extended Standard Model Based on Geometry* (Blaha Reasearch, Auburn, NH, 2013).

_____, 2014a, *Universes and Megaverses: From a New Standard Model to a Physical Megaverse; The Big Bang; Our Sister Universe's Wormhole; Origin of the Cosmological Constant, Spatial Asymmetry of the Universe, and its Web of Galaxies; A Baryonic Field between Universes and Particles; Megaverse Extended Wheeler-DeWitt Equation* (Blaha Reasearch, Auburn, NH, 2014).

_____, 2014b, *All the Megaverse! Starships Exploring the Endless Universes of the Cosmos Using the Baryonic Force* (Blaha Research, Auburn, NH, 2014).

_____, 2014c, *All the Megaverse! II Between Megaverse Universes: Quantum Entanglement Explained by the Megaverse Coherent Baryonic Radiation Devices – PHASERs Neutron Star Megaverse Slingshot Dynamics Spiritual and UFO Events, and the Megaverse Microscopic Entry into the Megaverse* (Blaha Research, Auburn, NH, 2014).

_____, 2015a, *PHYSICS IS LOGIC PAINTED ON THE VOID: Origin of Bare Masses and The Standard Model in Logic, U(4) Origin of the Generations, Normal and Dark Baryonic Forces, Dark Matter, Dark Energy, The Big Bang, Complex General Relativity, A Megaverse of Universe Particles* (Blaha Research, Auburn, NH, 2015).

_____, 2015b, *PHYSICS IS LOGIC Part II: The Theory of Everything, The Megaverse Theory of Everything, U(4)⊗U(4) Grand Unified Theory (GUT), Inertial Mass = Gravitational Mass, Unified Extended Standard Model and a New Complex General Relativity with Higgs Particles, Generation Group Higgs Particles* (Blaha Research, Auburn, NH, 2015).

_____, 2015c, *The Origin of Higgs ("God") Particles and the Higgs Mechanism: Physics is Logic III, Beyond Higgs – A Revamped Theory With a Local Arrow of Time, The Theory of Everything Enhanced, Why Inertial Frames are Special, Universes of the Mind* (Blaha Research, Auburn, NH, 2015).

_____, 2015d, *The Origin of the Eight Coupling Constants of The Theory of Everything: U(8) Grand Unified Theory of Everything (GUTE), S^8 Coupling Constant Symmetry, Space-Time Dependent Coupling Constants, Big Bang Vacuum Coupling Constants, Physics is Logic IV* (Blaha Research, Auburn, NH, 2015).

_____, 2016a, *New Types of Dark Matter, Big Bang Equipartition, and A New U(4) Symmetry in the Theory of Everything: Equipartition Principle for Fermions, Matter is 83.33% Dark, Penetrating the Veil of the Big Bang, Explicit QFT Quark Confinement and Charmonium, Physics is Logic V* (Blaha Research, Auburn, NH, 2016).

_____, 2016b, *The Periodic Table of the 192 Quarks and Leptons in The Theory of Everything: The U(4) Layer Group, Physics is Logic VI* (Blaha Research, Auburn, NH, 2016).

_____, 2016c, *New Boson Quantum Field Theory, Dark Matter Dynamics, Dark Matter Fermion Layer Mixing, Genesis of Higgs Particles, New Layer Higgs Masses, Higgs Coupling Constants, Non-Abelian Higgs Gauge Fields, Physics is Logic VII* (Blaha Research, Auburn, NH, 2016).

_____, 2016d, *Unification of the Strong Interactions and Gravitation: Quark Confinement Linked to Modified Short-Distance Gravity; Physics is Logic VIII* (Blaha Research, Auburn, NH, 2016).

_____, 2016e, *MoND: Unification of the Strong Interactions and Gravitation II, Quark Confinement Linked to Large-Scale Gravity, Physics is Logic IX* (Blaha Research, Auburn, NH, 2016).

_____, 2016f, *CQ Mechanics: A Unification of Quantum & Classical Mechanics, Quantum/Semi-Classical Entanglement, Quantum/Classical Path Integrals, Quantum/Classical Chaos* (Blaha Research, Auburn, NH, 2016).

_____, 2016g, *GEMS: Unified Gravity, ElectroMagnetic and Strong Interactions: Manifest Quark Confinement, A Solution for the Proton Spin Puzzle, Modified Gravity on the Galactic Scale* (Pingree Hill Publishing, Auburn, NH, 2016).

_____, 2016h, *Unification of the Seven Boson Interactions based on the Riemann-Christoffel Curvature Tensor* (Pingree Hill Publishing, Auburn, NH, 2016).

_____, 2017a, *Unification of the Eleven Boson Interactions based on 'Rotations of Interactions'* (Pingree Hill Publishing, Auburn, NH, 2017).

_____, 2017b, *The Origin of Fermions and Bosons, and Their Unification* (Pingree Hill Publishing, Auburn, NH, 2017).

_____, 2017c, *Megaverse: The Universe of Universes* (Pingree Hill Publishing, Auburn, NH, 2017).

_____, 2017d, *SuperSymmetry and the Unified SuperStandard Model* (Pingree Hill Publishing, Auburn, NH, 2017).

_____, 2017e, *From Qubits to the Unified SuperStandard Model with Embedded SuperStrings: A Derivation* (Pingree Hill Publishing, Auburn, NH, 2017).

_____, 2017f, *The Unified SuperStandard Model in Our Universe and the Megaverse: Quarks, ... ,* (Pingree Hill Publishing, Auburn, NH, 2017).

_____, 2018a, *The Unified SuperStandard Model and the Megaverse SECOND EDITION A Deeper Theory based on a New Particle Functional Space that Explicates Quantum Entanglement Spookiness (Volume 1)* (Pingree Hill Publishing, Auburn, NH, 2018).

_____, 2018b, *Cosmos Creation: The Unified SuperStandard Model, Volume 2, SECOND EDITION* (Pingree Hill Publishing, Auburn, NH, 2018).

_____, 2018c, *God Theory (*Pingree Hill Publishing, Auburn, NH, 2018).

_____, 2018d, *Immortal Eye: God Theory: Second Edition* (Pingree Hill Publishing, Auburn, NH, 2018).

_____, 2018e, *Unification of God Theory and Unified SuperStandard Model THIRD EDITION* (Pingree Hill Publishing, Auburn, NH, 2018).

_____, 2019a, *Calculation of: QED α = 1/137, and Other Coupling Constants of the Unified SuperStandard Theory* (Pingree Hill Publishing, Auburn, NH, 2019).

_____, 2019b, *Coupling Constants of the Unified SuperStandard Theory SECOND EDITION* (Pingree Hill Publishing, Auburn, NH, 2019).

_____, 2019c, *New Hybrid Quantum Big_Bang–Megaverse_Driven Universe with a Finite Big Bang and an Increasing Hubble Constant* (Pingree Hill Publishing, Auburn, NH, 2019).

_____, 2019d, *The Universe, The Electron and The Vacuum* (Pingree Hill Publishing, Auburn, NH, 2019).

_____, 2019e, *Quantum Big Bang – Quantum Vacuum Universes (Particles)* (Pingree Hill Publishing, Auburn, NH, 2019).

_____, 2019f, *The Exact QED Calculation of the Fine Structure Constant Implies ALL 4D Universes have the Same Physics/Life Prospects* (Pingree Hill Publishing, Auburn, NH, 2019).

_____, 2019g, *Unified SuperStandard Theory and the SuperUniverse Model: The Foundation of Science* (Pingree Hill Publishing, Auburn, NH, 2019).

_____, 2020a, *Quaternion Unified SuperStandard Theory (The QUeST) and Megaverse Octonion SuperStandard Theory (MOST)* (Pingree Hill Publishing, Auburn, NH, 2020).

_____, 2020b, *United Universes Quaternion Universe - Octonion Megaverse* (Pingree Hill Publishing, Auburn, NH, 2020).

_____, 2020c, *Unified SuperStandard Theories for Quaternion Universes & The Octonion Megaverse* (Pingree Hill Publishing, Auburn, NH, 2020).

_____, 2020d, *The Essence of Eternity: Quaternion & Octonion SuperStandard Theories* (Pingree Hill Publishing, Auburn, NH, 2020).

_____, 2020e, *The Essence of Eternity II* (Pingree Hill Publishing, Auburn, NH, 2020).

_____, 2020f, *A Very Conscious Universe* (Pingree Hill Publishing, Auburn, NH, 2020).

_____, 2020g, *Hypercomplex Universe* (Pingree Hill Publishing, Auburn, NH, 2020).

_____, 2020h, *Beneath the Quaternion Universe* (Pingree Hill Publishing, Auburn, NH, 2020).

_____, 2020i, *Why is the Universe Real? From Quaternion & Octonion to Real Coordinates* (Pingree Hill Publishing, Auburn, NH, 2020).

_____, 2020j, *The Origin of Universes: of Quaternion Unified SuperStandard Theory (QUeST); and of the Octonion Megaverse (UTMOST)* (Pingree Hill Publishing, Auburn, NH, 2020).

_____, 2020k, *The Seven Spaces of Creation: Octonion Cosmology* (Pingree Hill Publishing, Auburn, NH, 2020).

_____, 2020l, *From Octonion Cosmology to the Unified SuperStandard Theory of Particles* (Pingree Hill Publishing, Auburn, NH, 2020).

_____, 2021a, *Pioneering the Cosmos* (Pingree Hill Publishing, Auburn, NH, 2021).

Eddington, A. S., 1952, *The Mathematical Theory of Relativity* (Cambridge University Press, Cambridge, U.K., 1952).

Fant, Karl M., 2005, *Logically Determined Design: Clockless System Design With NULL Convention Logic* (John Wiley and Sons, Hoboken, NJ, 2005).

Feinberg, G. and Shapiro, R., 1980, *Life Beyond Earth: The Intelligent Earthlings Guide to Life in the Universe* (William Morrow and Company, New York, 1980).

Gelfand, I. M., Fomin, S. V., Silverman, R. A. (tr), 2000, *Calculus of Variations* (Dover Publications, Mineola, NY, 2000).

Giaquinta, M., Modica, G., Souchek, J., 1998, *Cartesian Coordinates in the Calculus of Variations* Volumes I and II (Springer-Verlag, New York, 1998).

Giaquinta, M., Hildebrandt, S., 1996, *Calculus of Variations* Volumes I and II (Springer-Verlag, New York, 1996).

Gradshteyn, I. S. and Ryzhik, I. M., 1965, *Table of Integrals, Series, and Products* (Academic Press, New York, 1965).

Heitler, W., 1954, *The Quantum Theory of Radiation* (Claendon Press, Oxford, UK, 1954).

Huang, Kerson, 1992, *Quarks, Leptons & Gauge Fields 2nd Edition* (World Scientific Publishing Company, Singapore, 1992).

Jost, J., Li-Jost, X., 1998, *Calculus of Variations* (Cambridge University Press, New York, 1998).

Kaku, Michio, 1993, *Quantum Field Theory*, (Oxford University Press, New York, 1993).

Kirk, G. S. and Raven, J. E., 1962, *The Presocratic Philosophers* (Cambridge University Press, New York, 1962).

Landau, L. D. and Lifshitz, E. M., 1987, *Fluid Mechanics 2nd Edition*, (Pergamon Press, Elmsford, NY, 1987).

Misner, C. W., Thorne, K. S., and Wheeler, J. A., 1973, *Gravitation* (W. H. Freeman, New York, 1973).

Rescher, N., 1967, *The Philosophy of Leibniz* (Prentice-Hall, Englewood Cliffs, NJ, 1967).

Rieffel, Eleanor and Polak, Wolfgang, 2014, *Quantum Computing* (MIT Press, Cambridge, MA, 2014).

Riesz, Frigyes and Sz.-Nagy, Béla, 1990, *Functional Analysis* (Dover Publications, New York, 1990).

Sagan, H., 1993, *Introduction to the Calculus of Variations* (Dover Publications, Mineola, NY, 1993).

Sakurai, J. J., 1964, *Invariance Principles and Elementary Particles* (Princeton University Press, Princeton, NJ, 1964).

Streater, R. F. and Wightman, A. S., 2000, *PCT, Spin, Statistics, and All That* (Princeton University Press, Princeton, NJ 2000).

Weinberg, S., 1972, *Gravitation and Cosmology* (John Wiley and Sons, New York, 1972).

Weinberg, S., 1995, *The Quantum Theory of Fields Volume I* (Cambridge University Press, New York, 1995).

Weinberg, S., 2000, *The Quantum Theory of Fields Volume III Supersymmetry* (Cambridge University Press, New York, 2000).

Weyl, H., 1950, *Space, Time, Matter* (Dover, New York, 1950).

INDEX

About the Author

Stephen Blaha is a well-known Physicist and Man of Letters with interests in Science, Society and civilization, the Arts, and Technology. He had an Alfred P. Sloan Foundation scholarship in college. He received his Ph.D. in Physics from Rockefeller University. He has served on the faculties of several major universities. He was also a Member of the Technical Staff at Bell Laboratories, a manager at the Boston Globe Newspaper, a Director at Wang Laboratories, and President of Blaha Software Inc. and of Janus Associates Inc. (NH).

Among other achievements he was a co-discoverer of the "r potential" for heavy quark binding developing the first (and still the only demonstrable) non-Aeolian gauge theory with an "r" potential; first suggested the existence of topological structures in superfluid He-3; first proposed Yang-Mills theories would appear in condensed matter phenomena with non-scalar order parameters; first developed a grammar-based formalism for quantum computers and applied it to elementary particle theories; first developed a new form of quantum field theory without divergences (thus solving a major 60 year old problem that enabled a unified theory of the Standard Model and Quantum Gravity without divergences to be developed); first developed a formulation of complex General Relativity based on analytic continuation from real space-time; first developed a generalized non-homogeneous Robertson-Walker metric that enabled a quantum theory of the Big Bang to be developed without singularities at t = 0; first generalized Cauchy's theorem and Gauss' theorem to complex, curved multi-dimensional spaces; received Honorable Mention in the Gravity Research Foundation Essay Competition in 1978; first developed a physically acceptable theory of faster-than-light particles; first derived a composition of extremums method in the Calculus of Variations; first quantitatively suggested that inflationary periods in the history of the universe were not needed; first proved Gödel's Theorem implies Nature must be quantum; provided a new alternative to the Higgs Mechanism, and Higgs particles, to generate masses; first showed how to resolve logical paradoxes including Gödel's Undecidability Theorem by developing Operator Logic and Quantum Operator Logic; first developed a quantitative harmonic oscillator-like model of the life cycle, and interactions, of civilizations; first showed how equations describing superorganisms also apply to civilizations. A recent book shows his theory applies successfully to the past 14 years of history and to *new* archaeological data on Andean and Mayan civilizations as well as Early Anatolian and Egyptian civilizations.

He first developed an axiomatic derivation of the form of The Standard Model from geometry – space-time properties – The Unified SuperStandard Model. It unifies all the known forces of Nature. It also has a Dark Matter sector that includes a Dark ElectroWeak sector with Dark doublets and Dark gauge interactions. It uses quantum coordinates to remove infinities that crop up in most

interacting quantum field theories and additionally to remove the infinities that appear in the Big Bang and generate inflationary growth of the universe. It shows gravity has a MOND-like form without sacrificing Newton's Laws. It relates the interactions of the MOND-like sector of gravity with the r-potential of Quark Confinement. The axioms of the theory lead to the question of their origin. We suggest in the preceding edition of this book it can be attributed to an entity with God-like properties. We explore these properties in "God Theory" and show they predict that the Cosmos exists forever although individual universes (or incarnations of our universe) "come and go." Several other important results emerge from God Theory such a functionally triune God. The Unified SuperStandard Theory has many other important parts described in the Current Edition of *The Unified SuperStandard Theory* and expanded in subsequent volumes.

Blaha has had a major impact on a succession of elementary particle theories: his Ph.D. thesis (1970), and papers, showed that quantum field theory calculations to all orders in ladder approximations could not give scaling deep inelastic electron-nucleon scattering. He later showed the eigenvalue equation for the fine structure constant α in Johnson-Baker-Willey QED had a zero at $\alpha = 1$ not 1/137 by solving the Schwinger-Dyson equations to all orders in an approximation that agreed with exact results to 4th order in α thus ending interest in this theory. In 1979 at Prof. Ken Johnson's (MIT) suggestion he calculated the proton-neutron mass difference in the MIT bag model and found the result had the wrong sign reducing interest in the bag model. These results all appear in Physical Review papers. In the 2000's he repeatedly pointed out the shortcomings of SuperString theory and showed that The Standard Model's form could be derived from space-time geometry by an extension of Lorentz transformations to faster than light transformations. This deeper space-time basis greatly increases the possibility that it is part of THE fundamental theory. Recently, Blaha showed that the Weak interactions differed significantly from the Strong, electromagnetic and gravitation interactions in important respects while these interactions had similar features, and suggested that ElectroWeak theory, which is essentially a glued union of the Weak interactions and Electromagnetism, possibly modulo unknown Higgs particle features, be replaced by a unified theory of the other interactions combined with a stand-alone Weak interaction theory. Blaha also showed that, if Charmonium calculations are taken seriously, the Strong interaction coupling constant is only a factor of five larger than the electromagnetic coupling constant, and thus Strong interaction perturbation theory would make sense and yield physically meaningful results.

In graduate school (1965-71) he wrote substantial papers in elementary particles and group theory: The Inelastic E- P Structure Functions in a Gluon Model. Phys. Lett. B40:501-502,1972; Deep-Inelastic E-P Structure Functions In A Ladder Model With Spin 1/2 Nucleons, Phys.Rev. D3:510-523,1971; Continuum Contributions To The Pion Radius, Phys. Rev. 178:2167-2169,1969; Character Analysis of U(N) and SU(N), J. Math. Phys. <u>10</u>, 2156 (1969); and The Calculation of the Irreducible Characters of the Symmetric Group in Terms of the

Compound Characters, (Published as Blaha's Lemma in D. E. Knuth's book: *The Art of Computer Programming Vols. 1 – 4*).

In the early 1980's Blaha was also a pioneer in the development of UNIX for financial, scientific and Internet applications: benchmarked UNIX versions showing that block size was critical for UNIX performance, developing financial modeling software, starting database benchmarking comparison studies, developing Internet-like UNIX networking (1982) and developing a hybrid shell programming technique (1982) that was a precursor to the PERL programming language. He was also the manager of the AT&T ten-year future products development database. His work helped lead to commercial UNIX on computers such as Sun Micros, IBM AIX minis, and Apple computers.

In the 1980's he pioneered the development of PC Desktop Publishing on laser printers and was nominated for three "Awards for Technical Excellence" in 1987 by PC Magazine for PC software products that he designed and developed.

Recently he has developed a theory of Megaverses – actual universes of which our universe is one – with quantum particle-like properties based on the Wheeler-DeWitt equation of Quantum Gravity. He has developed a theory of a baryonic force, which had been conjectured many years ago, and estimated the strength of the force based on discrepancies in measurements of the gravitational constant G. This force, operative in D-dimensional space, can be used to escape from our universe in "uniships" which are the equivalent of the faster-than-light starships proposed in the author's earlier books. Thus travel to other universes, as well as to other stars is possible.

Blaha also considered the complexified Wheeler-DeWitt equation and showed that its limitation to real-valued coordinates and metrics generated a Cosmological Constant in the Einstein equations.

The author has also recently written a series of books on the serious problems of the United States and their solution as well as a book on the decline of Mankind that will follow from current social and genetic trends in Mankind.

In the past twenty years Dr. Blaha has written over 80 books on a wide range of topics. Some recent major works are: *From Asynchronous Logic to The Standard Model to Superflight to the Stars, All the Universe!, SuperCivilizations: Civilizations as Superorganisms, America's Future: an Islamic Surge, ISIS, al Qaeda, World Epidemics, Ukraine, Russia-China Pact, US Leadership Crisis, The Rises and Falls of Man – Destiny – 3000 AD: New Support for a Superorganism MACRO-THEORY of CIVILIZATIONS From CURRENT WORLD TRENDS and NEW Peruvian, Pre-Mayan, Mayan, Anatolian, and Early Egyptian Data, with a Projection to 3000 AD*, and *Mankind in Decline: Genetic Disasters, Human-Animal Hybrids, Overpopulation, Pollution, Global Warming, Food and Water Shortages, Desertification, Poverty, Rising Violence, Genocide, Epidemics, Wars, Leadership Failure*.

He has taught approximately 4,000 students in undergraduate, graduate, and postgraduate corporate education courses primarily in major universities, and large companies and government agencies.

Recently he developed a quantum theory, The Unified SuperStandard Theory (UST), which describes elementary particles in detail without the difficulties of conventional quantum field theory. He found that the internal symmetries of this theory could be exactly derived from an octonion theory called QUeST. He further found that another octonion theory (UTMOST) describes the Megaverse. It can hold QUeST universes such as our own universe. It has an internal symmetry structure which is a superset of the QUeST internal symmetries.

www.ingramcontent.com/pod-product-compliance
Lightning Source LLC
Chambersburg PA
CBHW082009190326
41458CB00010B/3125